Evolution and the Foundations of Ethics

Evolution and the Foundations of Ethics

Evolutionary Perspectives on Contemporary Normative and Metaethical Theories

John Mizzoni

LEXINGTON BOOKS
Lanham • Boulder • New York • London

Published by Lexington Books
An imprint of The Rowman & Littlefield Publishing Group, Inc.
4501 Forbes Boulevard, Suite 200, Lanham, Maryland 20706
www.rowman.com

Unit A, Whitacre Mews, 26-34 Stannary Street, London SE11 4AB

British Library Cataloguing in Publication Information Available

Library of Congress Cataloging-in-Publication Data Available

ISBN 978-0-7391-9983-1 (cloth : alk. paper)
ISBN 978-0-7391-9984-8 (electronic)

∞™ The paper used in this publication meets the minimum requirements of American National Standard for Information Sciences—Permanence of Paper for Printed Library Materials, ANSI/NISO Z39.48-1992.

Printed in the United States of America

Contents

Chapter 1

Introduction

1.1 ANIMATING QUESTIONS AND STRATEGY FOR ANSWERING THEM

When we look at ethics from an evolutionary perspective, do we end up with a "survival of the fittest" ethic? With an evolutionary ethic, is eating your companion not necessarily wrong, in particular situations? In everyday conversation, the word "Darwinian" is often understood to mean a "competitive, do whatever is necessary, survival of the fittest" approach to life. These are misconceptions, however. This book, as a survey of ethical theories viewed from the perspective of evolution, aims to clear up these and related misconceptions about evolution and ethics.

In asking the question, "what does evolution imply about ethics?" we must be clear about what we mean by ethics. Do we mean ethics in a practical sense, such as ethical principles that can help us decide on a course of conduct? If so, then this is ethics understood at a *normative* level. *Normative ethical theories* focus on ethics in this sense, and they recommend and advise particular conduct. If we mean ethics in a more general sense, regarding the ultimate nature and status of ethics, this is ethics at a *metaethical* level. *Metaethical theories* focus on ethics in this ultimate sense.

In this work, I look at what several different models of evolution imply about four metaethical theories and six normative ethical theories. A full reckoning of ethics and evolution demands that we consider a full range of both metaethical and normative ethical elements.

More specifically, this book addresses the following big questions: If human biological evolution is part of our worldview, then how do commonplace notions of ethics fit in? And how do standard ethical theories fit into an evolutionary framework that sees human nature as the product of evolution? Do traditional ethical concepts and principles make sense when placed in an evolutionary framework? Or in the face of human biological evolution, should we abandon traditional ethical theories and systems? Do we need to create a new "evolutionary ethics" in place of more traditional ethical theories?

These philosophical questions constitute what I call "the philosophical problem of the nature and status of ethics in the light of evolution." In general, I characterize a philosophical problem as a cluster of interrelated philosophical questions. In an effort to answer these questions and develop a solution to this philosophical problem, in this book I lay out the main points of a wide assortment of normative and metaethical theories; then I consider what evolution implies about ethics, as characterized by these theories. The territory covered in this book is a range of standard metaethical and normative ethical theories, and standard evolutionary theory (for the most part).

Works dealing with evolution and ethics, works in so-called "evolutionary ethics," usually deal with the question "how did morality evolve?" or, "does evolution justify ethics?" or "can we prove that ethics comes from evolutionary processes?" T.H. Huxley (1825–1895), for instance, says he was responding to the following question, which is related, yet much different: whether there is a "sanction for morality in the ways of the cosmos" (Huxley 1894: 115). Huxley answered the question by arguing that there is no "sanction for morality" in the cosmic process, and that ethical progress in society depends on combating the cosmic process.[1] Yet a different question, "should the theory of evolution be adopted as an ethical principle?" is wrong-headed, because the theory of evolution at its heart is a scientific theory, not an ethical theory.[2] Although these are interesting and exciting questions, for me they are not the most basic questions about the topic. A more basic question would be to ask: Given what we know of ethics (through its most prominent metaethical and normative theories), how does evolution fit into the picture? If we understand human nature as the product of evolution, then what does this fact imply about these ethical theories? These more basic questions are my focus.

There are some books that discuss the evolutionary origins of morality; but for purposes of getting a sense of how ethics fits with an evolutionary view, they often do not directly engage the main ethical theories but only focus on a few key ethical concepts, such as altruism. *Evolution and the Foundations of Ethics* aims to be comprehensive and well-grounded in standard ethical and evolutionary theory, while at the same time being accessible to a wide and interdisciplinary audience.

1.2 BEYOND ALTRUISM

In setting out to answer how the conventional notions of ethics (those that are taught in most ethics classes) fit with evolutionary theory, I will not chiefly be concerned with *how* morality evolved. I will leave it to the evolutionary biologists to determine the evolutionary *mechanism* that gives rise to moral behavior in human beings. There has been much research and writing done that focuses on how morality as a human trait has evolved and even *could* evolve. Traditionally, for biological theorists there is an initial puzzle when we begin to ask *how* evolution can give rise to ethical behavior. The issue is in reconciling altruistic behavior with evolution. If an organism behaves extremely altruistically by sacrificing its life for another, and if that altruistic trait is always possessed by an organism that gives up its life, then how could it pass on its genes to the next generation? Over time, wouldn't that trait be naturally selected out? Darwin himself recognizes the issue and tries to address it (Darwin 1871: 163). When writers on evolution focus on reconciling altruistic behavior with evolution, and then venture into ethical territory, they tend to narrowly isolate *altruism* as the key (and only!) concept relevant to the discussion. A full reckoning of ethics and evolution, though, demands that we consider a wide *range* of ethical elements, not just one element. The paradox or problem of altruism, which is principally a question about the mechanism of evolution, is not my motivating philosophical problem.[3] As I have said, in this book I am not so concerned with the mechanism of evolution, but with how traditional ethical notions fit with an evolutionary view of human beings.

Further, for this project I do not assume that ethical behavior and altruistic behavior are one and the same, that ethical behavior *is* altruistic behavior. The study of ethics is an extremely broad enterprise and when studying ethical theory one inevitably comes across ethical *egoism* as a normative ethical theory. Although we might normally think of egoism as the *opposite* of ethical behavior, *ethical* egoism is indeed a normative ethical theory that comes in several different formulations.[4] A comprehensive study, which focuses on living traditions of ethics, must consider ethical egoism.

This book takes a broad approach. It seeks to bring together the common and salient normative factors that appear in different times, places, and cultures, and the major theories that have attempted to systematize those factors, and asks how these seemingly disparate and pluralistic ethical concepts, principles, theories, traditions, and practices can fit with evolutionary history. Commonsense morality, as understood by ethical theory, is my starting point. Part of my focus is on how to reconcile and accommodate the elements of "common sense morality" with an evolutionary framework; this is most apparent in Part II on normative ethics. Another part of my focus, which is most apparent in Part I on metaethics, is on how to reconcile and accommodate metaethical theories—theories *about* "common-sense morality"—with an evolutionary framework.[5]

1.3 MODELS OF EVOLUTION

In discussing evolution and ethics, one important point about evolutionary mechanism that is inescapable is how the process of biological evolution itself should properly be characterized. To begin to answer the question of how traditional ethical concepts and principles make sense when placed in an evolutionary framework, I will need a working model of human biological evolution. Having a particular model of evolution in mind will provide the focus needed to view traditional ethical concepts and principles in the light of evolutionary considerations.

In addition to considering Darwin's original view of evolution by natural selection, I will use several more recent and influential characterizations of the process of evolution as benchmarks against which to think through ethical concepts, principles, theories, and traditions. Beyond Darwin, the three principal theorists I will consider are zoologist Richard Dawkins (1979, 1982, 1989, and 1995), paleontologist Stephen J. Gould (2002), and theologian John Haught (2000, 2003, 2006, 2008, and 2010). The first two characterizations (Dawkins' and Gould's) are naturalistic, while the third is theistic (Haught's). Using these specific characterizations of ethics has upsides and downsides. An upside is that in discussing ethics I will not be referring to a general notion of "evolution" and leaving it up to the reader to interpret the sense of evolution I am talking about. A downside is that there may be other worthy and important characterizations of the biological process of evolution, and my work is somewhat limited by only focusing on these four.[6] Nevertheless, this method will allow us to make cross-comparisons of how ethics fits with different evolutionary theories, and develop a solution to the philosophical problem at hand.

According to Dawkins' model of biological evolution, evolution proceeds through gene selection; he disagrees with the classical Darwinian view of natural selection that views natural selection as selecting for individuals. Among the early competing evolutionary theories of the nineteenth century, classical Darwinism rejected the idea that species, communities, and ecosystems have adaptations that exist for their own benefit and instead claimed that natural selection selects for individuals. Dawkins, though, rejects the idea that *individuals* have adaptations that exist for their own benefit. For Dawkins, the unit of selection is the gene. Natural selection selects not for individuals, not for groups, but for genes.

> [T]he fundamental unit of selection, and therefore of self-interest, is not the species, nor the group, nor even, strictly, the individual. It is the gene, the unit of heredity. (Dawkins 1989: 11)

Instead of conceptualizing the process of evolution as organisms and species mercilessly competing against each other for resources, which results in a prolific environment filled with efficient and well-organized relationships between organisms, and between organisms and their environment, for Dawkins we should think in terms of genes competing against each other. Genes compete with other genes and a feature of the history of life is that many genes that have built vehicles for themselves have done very well in propagating themselves. These vehicles may be individual organisms, or groups.

A related concept distinctive of Dawkins' model of evolution is the concept of an extended phenotype. In addition to genes building bodies they are physically connected to, genes can also build, influence, or manipulate other bodies—inanimate or animate—that they can use to help themselves survive. Examples include: the houses that caddis flies build, the dams that beavers build, webs that spiders spin, and mounds that termites build. These "creative" behaviors are genetically based and they enable genes to persist through geologic time. Dawkins also points to situations in which genes share a given extended phenotypic trait; for example, both fluke genes and snail genes can influence the thickness of a snail's shell (Dawkins 1982: 200, 209–11).

Another important aspect of Dawkins' characterization of evolution is extrapolationism (Sterelny 2001: 132). As Darwin did, Dawkins views evolution as a long, slow, accumulative process; and further, like Darwin, Dawkins views evolutionary patterns as accumulations of microevolutionary events over vast stretches of time.

An aspect of Dawkins' characterization of evolution, one that follows from his genic selection—and highly relevant to evolutionary ethics—is his position that evolutionary theories of social behavior apply to *human* behavior. Dawkins favors genic explanations of human cooperation and altruism. Kin selection explains how humans are cooperative with their relatives (Dawkins 1989, chapter 6), and reciprocal altruism explains how humans are cooperative with strangers (Dawkins 1989, chapter 10; 1982: 153–55). Both kin selection and reciprocal altruism help to solve the biological puzzle of evolution and altruistic behavior (§1.2, above).

Like Dawkins, Stephen Jay Gould has developed a naturalistic theory of evolution. Gould sums up his twenty-five years of writing on evolutionary theory in his monumental work *The Structure of Evolutionary Theory* (2002). Gould maintains that the essence of Darwinian theory (natural selection) can be seen as a tripod, since the theory has three central principles: agency, efficacy, scope (Gould 2002: 11, 14–15). Agency, the first part of the tripod, concerns the units of selection. As noted above, Darwin held that individual organisms are the units of selection. Darwin believed that "natural selection works on *organisms* engaged in a struggle for personal success, as assessed by the differential production of surviving offspring" (Gould 2002: 126). Efficacy, the second part of the tripod, concerns the relative power of natural selection to drive evolution. While most of Darwin's contemporaries held that natural selection would be a force too weak to account for the evolution of novelties in nature, Darwin believed that natural selection was indeed sufficiently efficacious to remove the unfit as well as to be "the positive mechanism of evolutionary novelty" (Gould 2002: 14). Darwin believed that natural selection, the principal though not the *exclusive* cause of evolutionary change, could craft new characteristics of organisms (adaptations) (Gould 2002: 147, 254).

Scope, the third leg of the Darwinian tripod, makes the claim that natural selection as a microevolutionary process proceeding gradually over millions of years creates all microevolutionary *and* macroevolutionary phenomena. Darwin believed that natural selection could "produce the full panoply of taxonomic diversity and apparent 'progress' in complexification of morphology through geological time" (Gould 2002: 15).

Gould believes that a century's worth of evolutionary biology has modified the original three legs of the essential Darwinian theory. The first leg (agency) has been modified by many different evolutionists, including genic selectionists such as Dawkins. Genic selectionists have departed from Darwin's view that individual organisms are the units of selection. Gould, in explaining his revised Darwinian model, specifies what an individual needs in order to qualify as a unit of selection: a material entity must have certain characteristics such as a discrete and definable birth and death, and sufficient stability during its lifetime (Gould 2002: 602). Using this definition of an individual, Gould believes there are many kinds of individuals that could play the role of an evolutionary individual, that is, a unit of selection.

> [T]hese principled exclusions leave us with a rich hierarchy of legitimate biological individuals. . . . [I]ndividuals at many levels—including genes, cell lineages, organisms, demes, species, and clades—can act as units of Darwinian selection. (Gould 2002: 613)

Whereas both Darwin and Dawkins espouse unilevel theories of selection (Darwin says natural selection selects for organisms; Dawkins says natural selection selects for genes), Gould holds a *multilevel* theory of selection; he holds that natural selection acts at multiple levels. He calls his view of selection the hierarchical theory of selection (Gould 2002: 31).[7] In his work as a paleontologist, Gould, like his paleontological colleagues, initially viewed the pattern in the fossil record of stasis and then abrupt appearance of fossil species as due to an imperfect fossil record. But he and one of his paleontologist colleagues, Niles Eldredge, developed a theory of punctuated equilibrium that interprets the long periods of no change at the species level, which they call stasis, that are then interrupted by rapid speciation, as the actual path that macroevolution would naturally take. In their view, the fossil record was commonly seen as highly imperfect because of the widespread assumption that macroevolution proceeds continually, slowly, and gradually over millions of years. If we give up the assumption of gradualism, they contend, and instead accept a theory of punctuated equilibrium, then a fossil record lacking evidence of gradual speciation over geological time is what one would expect.

For Gould, species have the characteristics that are necessary for them to be viewed as units of selection. Species have clearly definable births and deaths and are sufficiently stable during their "lifetimes": "the great majority of species," writes Gould "remain essentially stable throughout their geological lifetimes" (Gould 2002: 606). The theory of punctuated equilibrium encourages us to look at species as individuals; thus, the theory provides support for Gould's hierarchical selection that says that natural selection occurs at many levels not only one.

Gould maintains that the second leg of the Darwinian tripod (efficacy) has also been modified. He accepts that natural selection is efficacious as a creative biological process, but he argues that there are other internal, formal, and structural constraints at work in the process. Gould's model

> introduces a sufficient weight of formalist thinking—*via* renewed appreciation for the enormous importance of structural, historical, and developmental constraint in channeling the pathways of evolution, often in highly positive ways. . . . [A] strictly Darwinian . . . approach to adaptation no longer suffices to explain the channeling of phyletic directions, and the clumping and inhomogeneous population of organic morphospace. (Gould 2002: 21)

In Gould's model of evolution, the third leg of the Darwinian tripod (scope) has also been modified. Gould agrees with Darwin that natural selection as a *micro*evolutionary process proceeds gradually over millions of years, but Gould does not think the microevolutionary process can adequately explain all *macro*evolutionary phenomena. He thus disagrees with extrapolationism, the traditional Darwinian view that microevolutionary processes can fully account for all macroevolutionary phenomena. While Darwin and Dawkins hold that extrapolationism is a sound working theory, Gould does not (Sterelny 2001: 132–33).

With the Darwinian tripod thus modified, we now have

> the hierarchical theory of selection on leg one, the structuralist critique of Darwinian functionalism and adaptationism on leg two, and the paleontologist's conviction (leg three) that general macroevolutionary processes and mechanisms cannot be fully elucidated by uniformitarian extrapolation from the smallest scale of our experiments and personal observations. (Gould 2002: 970)

Gould sees several problems with Dawkins' model of evolution. Gould argues that Dawkins does not appreciate that a unit of selection—an evolutionary individual—must interact with the environment. Genes do not do this, Gould says; genes replicate. Gould thus thinks that genes do not play a causal role in natural selection, but rather they merely record organic changes. Gould therefore claims that genic selection commits a causal fallacy.

> [W]hile genes may be appropriately designated as fundamental replicators . . ., replicators simply aren't units of selection or, for that matter, causal agents at all under our usual notions of mechanism in science. The misidentification of replicators as causal agents of selection—the foundation of the gene-centered approach—rests upon a logical error best characterized as a confusion of bookkeeping with causality. (Gould 2002: 614)

Gould believes that Dawkins' notion of *extended phenotype* is also a mistake on Dawkins' part. He views Dawkins' concept of an extended phenotype as a subtle reversion to the classical Darwinian view that the organism is the unit of selection, which is the view that Dawkins' gene-centered approach was intended to replace (Gould 2002: 638–41).

Gould and Dawkins also disagree on the importance of punctuated equilibrium. Gould believes that punctuated equilibrium unites the three legs of his revised Darwinian tripod (Gould 2002: 970–71). Dawkins does agree that there may be "punctuations of rapid evolution, or even abrupt macromutation," and "sudden extinctions" (Dawkins 1995: 83). But Dawkins sees punctuated equilibrium as more of a "minor wrinkle on the surface of neo-Darwinian theory" (Dawkins 1987: 251).

Another area of contrast between the evolutionary models of Dawkins and Gould concerns human behavior. Dawkins is confident about applying evolutionary theories of social behavior to human behavior while Gould is skeptical of purely genetic explanations of human behavior and any claims that genes determine human behavior. Over the years, (going all the way back to the early days of evolutionary biology), many explanations of human behavior that have made use of the concepts and principles of evolutionary biology have been failures, thinks Gould (Gould 1981). In Gould's view, the human genetic constitution does predispose us to certain behaviors, yet compared with the influence of culture, the genetic influence is actually weaker. We are very ignorant, Gould says, when it comes to the biology of human behavior. He is well-known for closely scrutinizing and critiquing various attempts by sociobiologists to draw conclusions about specific human behaviors from evolutionary models (Kitcher 2004: 4).

Turning now to a view of evolution that is theistic instead of naturalistic, let us consider the model of evolution that John Haught has defended in several recent books: *God After Darwin* (2000), *Deeper than Darwin* (2003), and *Is Nature Enough?* (2006). Theistic evolutionary theories are not new. In Darwin's time, theorists such as Asa Gray and Alfred Russell Wallace held theistic evolutionary theories; they both retained a supernaturalist aspect in their evolutionary theories because of what they understood as the inability of a naturalistic theory of evolution to account for the human mind.[8] Haught's work provides a recent example of a sustained effort to work out a theistic model of evolution. As an expert in this area, Haught was called to testify in the 2005 court case in Pennsylvania (*Kitzmiller et al. v. Dover District School Board*), in which a suit was brought against the Dover School Board because it proposed to incorporate "intelligent design" into high school biology classes (Haught 2008: xi). Haught testified that teaching intelligent design, as a religious notion, is inappropriate for biology classes. The court ultimately ruled that teaching intelligent design in science classes would be a violation of the First Amendment's Establishment Clause (Haught 2008: 194).

For Haught, God is the ultimate explanation of evolution and the cosmic process (Haught 2000: 165). He believes that there is a way of reading evolution that is consistent with science *and* religious hope (Haught 2003: 25). Although Dawkins and Gould espouse metaphysical naturalism (or materialism), Haught says *their* metaphysical commitments should not lead us to suppose that an *evolutionary* understanding requires a commitment to metaphysical naturalism (or materialism). For, Haught claims, "There is absolutely no reason why evolutionary science cannot be methodologically naturalistic—following the proper decision of all science to leave out any appeal to nonnatural explanations—without being metaphysically materialistic" (Haught 2000: 32).

Haught's approach is to assent to *methodological* naturalism and concede from the outset the general integrity of Darwinian science (Haught 2000: 14; 2003: xi). For Haught's theory of theistic evolution, the differences between genic selection or individual selection or group selection are scientific disagreements that do not make a difference to his overall model of evolution. He says that a precise formulation of the nature of selection is strictly an issue for biological science because a theological explanation cannot contribute to scientific explanations. Thus, whether the history and process of evolution is described more appropriately with an extrapolationist account or with an account of punctuated equilibrium, it would not make a difference for him either.[9] Haught accepts the disorderly, undirected aspects of evolution that evolutionists say are part of the life process; and he says these are features of natural history that proponents of intelligent design theory pass over (Haught 2000: 4).

> Nature is ordered, yet open to disorder . . . Contingency doesn't prevent order from emerging. Natural selection is rigorously lawful, yet open to indeterminate new creation. The world has a temporal character that allows evolution to occur. (Haught 2000: 96)

Haught concurs with Gould that the unfolding of natural history is not

> linearly progressive. For vast periods of time little happened, and much of the history of life's evolution can be captured in the image of a randomly branching bush. (Haught 2000: 117)

Haught accepts that the cosmos is an unfinished process, since there is undeniable evidence that the universe is *still* being created and is not an orderly design (Haught 2000: 6). He sees that nature itself is inherently historical, and the cosmos too has always had a historical character, given the "big bang" cosmology to which most scientists today subscribe (Haught 2000: 156).

That the evolution of life on earth is a very long story is extremely important for Haught. The details that scientists are actively engaged in working out, in his view, cannot undermine the fact that the universe is indeed one long story, one that is continuing to unfold.[10] He calls the fact that the universe is indeed one long story the narrative cosmological principle (Haught 2003: 61). According to Haught:

> It could be argued that because of recent science's discovery of nature's fundamentally historical character, we may with more confidence than ever locate *the whole cosmos*, and not just human affairs, within the horizon of the promise that molds the experience of Israel, Jesus, and his followers. (Haught 2000: 156–57)

Haught views the unfolding universe as one long experiment in self-actualization. The "experimental self-actualizing" of the universe takes deep time in order to play itself out (Haught 2006: 190).

Haught denies that naturalistic evolutionists have told the *whole* story about how life evolves; he calls it a reading problem (Haught 2003: chapter 2).

Taking things too literally, as he sees it, affects not only religious fundamentalists but scientists as well (Haught 2003: xv). As Haught puts it: "Some evolutionists think that naturalism can give an ultimate explanation of everything" (Haught 2006: 15). In contrast, Haught describes different ways of reading reality, because solely relying on one way of reading reality, he says, will result in an incomplete account. To be fully explanatory, Haught says we ought to employ many different frameworks, only one of which is scientific. To presume that only science can make true claims about reality is restrictive and reductive in his view. To explain his position Haught uses the notion of a layered explanation:

> Suppose that a wood fire is burning in your backyard. Your neighbour comes over and asks you to explain why the fire is burning. A very good response would be: it is burning because the carbon in the wood is combining with oxygen to make carbon dioxide. This is an acceptable explanation, and for a certain kind of inquiry it is enough. Still, there can be other levels of explanation. For example, you might just as easily have answered your neighbour's question by saying: the fire is burning because I lit a match to it.
>
> And a third answer might be: "The fire is burning because I want to roast marshmallows." Different levels of explanation, as is evident here, can coexist without conflict. "I want to roast marshmallows" does not in any way compete with physical explanations of the burning wood. I do not respond to my neighbour: "The fire is burning because of chemical combustion *rather than* because I want marshmallows." "I want to roast marshmallows," in fact, cannot be squeezed into the explanatory slot that focuses on the chemistry of combustion and my overarching purpose of wanting something to eat. (Haught 2006: 16)

In terms of the three different layers of explanations, this time in an example about water boiling on the stove because I want a cup of tea, Haught observes:

> Here we have three logically distinct explanations. All are correct and relevant, but they cannot be reduced to or mapped onto one another. Each adds something important to an understanding of why the water is boiling, and it does so without conflicting or competing with the others. (Haught 2006: 70)

Haught's layered explanations approach also applies to human behavior. With regard to evolutionary explanations of human behavior, Haught agrees with Gould that exclusively biological explanations come up short. Just as the story of life on earth and the origins of human critical intelligence are not wholly captured by scientific explanations, the biological sciences cannot fully explain and understand human behavior by reducing it to genetic explanations. For Haught, Darwinism does not tell us everything we need to know about life (in the biographical sense), or about living a life (again, biographically) (Haught 2003: xi).

1.4 "EVOLUTIONARY ETHICS"

The project of assessing how commonplace notions of ethics fit with an evolutionary framework is not an entirely new one. Since the mid-nineteenth century, evolutionists have sought to view ethics through an evolutionary prism. In chapter 4 of *The Descent of Man* (1871), Darwin himself ventured into this territory, and speculates about evolution and ethics.[11] Darwin's approach is rather well balanced, although not fully developed.

Other attempts to bring ethics and evolution together have unfortunately been notorious blunders. In the minds of some readers today, the very phrase "evolutionary ethics" conjures up these past blunders. The following is a simple nineteenth-century example of a clumsy union of evolution and ethics in the attempt to create an "evolutionary ethic." Upon analysis we can see it simply as a hurried fusion of utilitarian ethical theory with an evolutionary view. The first step is to accept the principle of utility—that we ought to bring about the greatest happiness for the greatest number—as our foundational ethical principle. Second, we notice that biological evolution proceeds by organisms and species mercilessly competing against each other for resources, and this has resulted in a prolific environment filled with efficient and well-organized relationships between organisms, and between organisms and their environment. From this we reach the conclusion that human beings should abide by the principle of survival of the fittest as a way to bring about the greatest good for the greatest number. Under this view, we should therefore allow and promote human beings to compete ruthlessly against each other for resources, all the while being careful not to aid the weakest and most vulnerable humans, for fear of disrupting the efficiency of the evolutionary process in yielding maximum results.[12] Such strategies for combining utilitarian ethics and evolution in order to fashion an "evolutionary ethic" are, as I said above, rather clumsy, and much too blunt. Unlike today's theorists who often narrowly focus on the concept of *altruism*, these early attempts at evolutionary ethics focused on one aspect of utilitarian ethics, namely *the ethical principle of utility*, or something close to it. In the chapters ahead, when I discuss an ethical theory in conjunction with evolutionary theory, I do not consider one concept or principle from that ethical theory, but I consider all of the most salient elements of an ethical theory. In my discussions of ethics and evolution, then, I am able to bring out more of the nuanced connections between evolution and ethics.

In order to make progress in the field of "evolutionary ethics," we need to be exceedingly careful about our treatment of evolution and ethics. To make progress in this area, we must struggle to break away from common misunderstandings both about biological evolution itself and about the interface of biological evolution and ethics.

From my own experience, I know that as soon as I say I am working with evolution and ethics, people will often assume that I am doing "evolutionary ethics." But what is "evolutionary ethics," really? The historian of science Paul Lawrence Farber (1999: 96), who has written a historical survey of such theories, says that "'evolutionary ethics' generally refers to attempts to use the theory of evolution as a foundation for ethics," and in such an endeavor "appeals to scientific foundations for justification are central."[13] The philosopher of science Michael Ruse says that evolutionary ethics is "the attempt to locate and ground morality in our biological origins" (Ruse 1999: 198). The philosopher Neil Levy defines evolutionary ethics as the attempt to prove that ethics comes from evolution (Levy 2002: 128). And some writers speak of a "system of evolutionary ethics" (Nitecki 1993: 3; Richards 1993: 122). Given these ways of characterizing "evolutionary ethics," it is still unclear, though, whether "evolutionary ethics" is meant as a metaethical or normative ethical theory, or both. Upon closer inspection, it turns out that "evolutionary ethics" is a hodgepodge of some metaethical and some normative strands, and is heavily dependent on the individual writer's interpretation. Nitecki, an editor of a volume of essays on "evolutionary ethics," claims that "evolutionary ethics" may be hard to define because of its interdisciplinary nature, and the subject unfortunately becomes "hazy" (Nitecki 1993: 16).

Nitecki is correct that "evolutionary ethics" is hazy, but I don't think it is because of its interdisciplinary nature, but simply that it is an ill-formed term. Consider a simple example of how things easily get off on the wrong foot when discussing evolution and ethics. R.J. Richards is a self-proclaimed evolutionary ethicist and his article "A Defense of Evolutionary Ethics" (1986), which was published in the inaugural issue of the journal *Biology and Philosophy*, has been reprinted in several anthologies.[14] Richards speaks of "ethics born of evolutionary facts," of ethics "founded on" evolution, of ethics "based on" evolution, as well as "Darwinizing in morals," and "systems of evolutionary ethics" (Richards 1993: 120, 122, 123, 124, 125). All of these are telling phrases of how Richards conceives of his project.

To make the point that "evolutionary ethics" is an ill-formed term, I will focus on one of Richards' remarks about Darwin. Richards tells us that Darwin's "earliest speculations on species change . . . also included hypotheses about an evolutionary ethics" (Richards 1993: 113). It is a seemingly innocuous statement; yet it is a way of framing the issue that quickly leads us down the wrong path. In the way Richards frames the issue, Darwin, in merely *thinking* about ethics from an evolutionary perspective, was thereby framing "hypotheses about an evolutionary ethics" and considering "an evolutionary theory of morality" (Richards 1993: 113). This way of framing the discussion immediately leads to confusion. For consider what earlier theorists were discussing when they were *not* viewing ethics from an evolutionary perspective. Shall we say they were framing hypotheses about "a theistic ethics"? That seems to make enough sense. Yet we can quickly see how this way of characterizing an ethical theory will tend to distort what we take as the subject matter at hand. Consider Plato. What was he doing when thinking about ethics? Was he framing hypotheses about a "Forms ethics"? Was Aristotle entertaining hypotheses about a "substance ethics"? And were Hume and Mill entertaining hypotheses about "phenomenalist ethics"? And Kant "things-in-themselves" ethics?

Such phrases are not used in moral philosophy, since they are not helpful ways of characterizing theories of moral philosophy. Theism, Forms, substance, thing-in-themselves, etc., are simply part of the background metaphysics of the person who is thinking about ethics. And the same is true with evolution. Normally, when discussing ethics, it is not enlightening to bring one's metaphysical commitments to the foreground. Of course, one's metaphysical commitments are important, and rather inescapable; yet they are not central. For example, it would be more revealing of one's ethic if one were to focus on the *ethical* dimensions of it. Take Virtue Ethics, for instance: it might be a theistic Virtue Ethics, or a Platonic Virtue Ethics, or an Aristotelian Virtue Ethics, or a Virtue Ethics à la David Hume. In all these cases it is an ethic that emphasizes virtues as most central. Its particular metaphysical background is significant in how the ethical approach is distinguished from other variants of virtue ethics, no doubt. Yet Richards and other proponents of an "evolutionary ethic" write as if specifying the prominent metaphysical ingredient in some way puts this kind of theory in a special league of some kind. It doesn't, though. And, in fact, Richards' highlighting the *evolutionary* component and hiding its true *ethical* component obscures discussion. It is fine to say that Darwin was thinking about evolution and ethics, but it is still unclear what kind of *ethic* is under discussion. Is it a normative ethics that prescribes particular conduct, on a plane with utilitarian ethics, virtue ethics, or deontological ethics? Or is it a metaethic concerning the ultimate nature of ethics, on a metaethical plane with moral realism, moral relativism, or expressivism?

Put more specifically, an "evolutionary ethic," by being "evolutionary" commits itself to the claim that human beings are the product of evolution, so it makes a claim about human nature. By and large, ethical theories *usually* make claims and assumptions about human nature. Theists make claims about human nature; Plato has them, as well as Aristotle, Hume,

Kant, Mill, among others. There is nothing special here. Along these lines, by designating a theory as an "evolutionary ethic," that would then mean that an "evolutionary ethic" is any theory that can accept that human beings are products of evolution. But today, we find proponents of the full range of ethical theories—from normative to metaethical (virtue ethics, deontological ethics, utilitarian ethics, moral realism, moral relativism, expressivism, etc.)—who, as we'll see in the chapters to follow, accept that human beings are products of evolution. That then makes the label "evolutionary ethics" basically vacuous, since each of these ethical theories counts as an "evolutionary ethic."

1.5 ETHICS AND ETHICAL THEORIES

The ethical concepts, principles, theories, and traditions I will consider in this book are the ones that have proven themselves to be useful and explanatory when trying to understand and evaluate a variety of ethical phenomena. Ethical phenomena range from ethical behavior, ethical motivation, ethical questions, ethical dilemmas, ethical issues, ethical discussions, ethical deliberations, ethical analyses, ethical positions, and ethical traditions. In this book, I do not offer evolutionary theory as the justification for these ethical concepts, principles, theories, or traditions; these ethical tools have already been justified by their *own* standards, on the playing field of ethical theory. I am also not trying to derive ethical norms from evolutionary theory. What I *am* doing is simply looking at a range of ethical tools alongside an evolutionary conception of human beings in order to see if they fit together, and if so, how do they fit, exactly?

Today, when discussing evolution and ethics, the most important and necessary clarification to observe is not the is-ought gap—the gap between statements of fact (is) and statements of ethics (ought)—as too many commentators suggest.[15] A more important clarification is the distinction between metaethics and normative ethics. When writing about "ethics" one must be clear whether one means normative ethics or metaethics, for they are two very different enterprises.

The ethical theories I discuss in this book have been developed by moral philosophers. With metaethical theories (e.g., Error Theory, Expressivism, Moral Relativism, and Moral Realism), moral philosophers attempt to explain the general nature and origins of ethics. The most basic metaethical questions are: What is morality, what is its status, and where does it come from? The first four chapters of this book (Part I) look at how the four different metaethical theories of error theory, expressivism, relativism, and realism fit with an evolutionary perspective.

Briefly, in terms of the *status* of morality, error theorists claim that there is a fundamental mistake, an error, at the heart of commonsense morality. Expressivists claim that moral statements and judgments are not the kinds of claims that can be true or false, since moral statements and judgments are expressions of emotion. Moral relativists claim that ethical statements and judgments *can* be true or false, and whether a particular ethical statement is true or false is always relative to a particular context (or framework, or society, or culture). Moral realists also claim that ethical statements can be true or false, but they contend that some ethical statements and judgments are objectively true, not only relatively true. Again, we can see that these metaethical theories are about the *status* of morality, not the particular directives of morality.

The six chapters that follow, which comprise Part II of the book, look at how normative ethical theories fit with evolution. The normative level of ethical theory attempts to answer practical questions like, how should I live? and, what is the right thing to do? Theorists who work with normative ethical theories (such as Virtue Ethics, Natural Law Ethics, Social Contract Ethics, Utilitarian Ethics, Deontological Ethics, and Ethics of Care) attempt to articulate and defend normative ethics that people can and do use in a practical way when deliberating about specific actions. Normative ethical theories are the kinds of ethical theories that provide guidance about what we ought to do (should I follow a natural law, work to bring about happiness for all concerned, or care for those who are close to me?), or guidance about what kind of persons we ought to become (what virtues are most important for me to develop?). Metaethical theories, as merely general theories about the nature, status, and origin of ethics, do not provide this kind of practical guidance. Part II (chapters 6 through 11) will be taken up with describing the various ways in which normative theories and their key concepts and principles mesh with what evolutionists would have us believe about the evolution of human beings. In the final chapter of the book (chapter 12), I will summarize my overall findings about evolution and ethics.

NOTES

1. (Huxley 1894: 134).
2. For a good introduction to the scientific theory of evolution, see Coyne (2009).
3. Of "the vexatious problem of altruism," Stephen Jay Gould says it was "previously the greatest stumbling block to a Darwinian theory of social behavior" (Gould 1980: 256).
4. For example, act-egoism, rule-egoism, and social contract ethics (see chapter 8).
5. See Gert 2004.

6. Considering a model of cultural evolution, such as Joseph Henrich's, might also be worthwhile for a project like mine.

7. Sober & Wilson (1998) also defend a multilevel model of natural selection.

8. The Catholic Church is situated in this tradition of theistic evolution. Pope John Paul II (1997) and Cardinal Christoph Schönborn (2005), for largely the same reasons offered by Gray and Wallace, maintain that a purely naturalistic theory of evolution cannot fully explain the human mind. Even Pope Pius XII (1950: 287) said that evolution is an open question, as long as it is confined to the development of the human body. See also the collection of essays in Clayton & Schloss (2004).

9. But many of his comments about the impotence of deep time to account for the ultimate emergence of critical intelligence does entail that he believes that mere extrapolationism cannot deliver an explanation for all evolutionary developments.

10. For Haught, the fact that the universe is still unfolding provides us with a basis for religious hope, for there is a way of reading evolution that is consistent with religious hope (2003: 25).

11. As Flew (1967: 1) points out, Herbert Spencer's work on evolution and ethics predates Darwin's. Nevertheless, though, Flew asserts that Darwin should be the reference point at which to begin. In each chapter below, I look closely at Darwin's views on ethics. Two other early examples are Thomas H. Huxley and John Dewey. In the lecture "Evolution and Ethics" (1894: 132–34) Huxley is critical of the evolutionary ethics of his day. He thinks that evolutionary ethicists have overlooked the fact that the cosmic process proceeds through natural selection, while ethical progress proceeds by "combating" the cosmic process (1894: 134). In the article "Evolution and Ethics" (1898: 109), Dewey says he disagrees with Huxley's "dualism between the cosmic and the ethical." Dewey attempts to frame a more nuanced relationship between evolutionary concepts and ethical concepts, and thinks it is unfortunate that evolutionary ethicists have not been as careful.

12. This kind of reasoning is usually attributed to Herbert Spencer, for example, (Abbott 2005: 178), but it is more of a simplified version of what Spencer had in mind (Rachels 1990: 62–66).

13. Farber's extended historical survey of evolutionary ethics is *The Temptations of Evolutionary Ethics* (1994).

14. Reprinted in (Thompson 1995: 249–91; Ruse 2009: 388–410; and Levy 2010).

15. It is not that I reject the idea of an is-ought gap. I think it does have serviceable, logical uses.

Part I

EVOLUTION AND METAETHICS

Chapter 2

Evolution and Error Theory

Several species of cuckoo birds will lay their eggs in the nests of a different bird species. Foster-parent birds are then "tricked" into caring for the cuckoo chicks. The foster-parent birds commit an "error" by extending their efforts and energies into caring for cuckoo chicks, because the cuckoos do not possess the foster-parent bird's genes (Dawkins 1989: 102). A rule built into the foster bird parent that says "Be nice to any small bird sitting in the nest that you built" prompts the foster-parent birds to commit this error. But the rule still serves the foster-parent bird's genes well enough for those genes to successfully compete in the struggle for genic survival in the long run, since the small bird sitting in the nest that the parent built will *usually* contain the parent bird's genes.

Is human morality a genetically implanted set of rules that amounts to an error, a trick of evolution?

Before we look at error theory as it is coupled with evolutionary considerations, we must first get clear on what error theorists claim about the nature of ethics. In (§1.5) I initially described a metaethical theory as a theory that attempts to explain the general nature and origins of ethics. At a minimum, a metaethical theory seeks to provide answers to the questions: What is morality, what is its status, and where does it come from? Concerning the general nature and origins of ethics, error theorists argue that there is a fundamental mistake, an error of some kind, at the heart of commonsense morality. In this chapter I first lay out the main points of error theory as they have been articulated by John Mackie in his influential *Ethics: Inventing Right and Wrong* (1977). Mackie is a philosopher who is well known for putting forward error theory as an account of the ultimate nature of ethics, and his main ideas include: commonsense morality, objective values, internalism, the argument from relativity, the argument from queerness, and patterns of objectification.

After outlining the basic mechanics of error theory, I then discuss how two recent proponents of error theory, Michael Ruse and Richard Joyce, have attempted to join ethical error theory with an evolutionary perspective. After that I examine what the models of evolution put forward by Darwin, Dawkins, Gould, and Haught imply about error theory in ethics. Finally, I conclude the chapter by summarizing my critical assessment of error theory and the attempt to support it with evolutionary theory.

2.1 MACKIE'S ERROR THEORY

When Mackie reflects on and examines commonsense morality, one of the most important aspects of it he finds is that it presupposes the existence of objective values. He writes,

> I conclude, then, that ordinary moral judgments include a claim to objectivity, an assumption that there are objective values. . . . And I do not think it is going too far to say that this assumption has been incorporated in the basic, conventional, meanings of moral terms. Any analysis of the meanings of moral terms which omits this claim to objective, intrinsic, prescriptivity is to that extent incomplete. (Mackie 1977: 35)

Mackie thus argues that one of the most significant features of commonsense morality that any theory of morality must reckon with is the assumption that some values are objective. In Mackie's estimation, commonsense morality simply assumes that there are objective values. The error that Mackie speaks of in his error theory concerns the status of these objective values. Having studied and investigated the ethical landscape, Mackie concludes that "there are no objective values" (Mackie 1977: 15). Although belief in objective values is an assumption of commonsense morality and has been a fixture in ordinary morality and ethical theory for centuries, using several influential arguments, Mackie mounts the case that commonsense morality is premised upon an error. His error theory proposes that

> although most people in making moral judgments implicitly claim, among other things, to be pointing to something objectively prescriptive, these claims are all false. (Mackie 1977: 35)

As Mackie describes it, espousing an error theory about ethics is

> admitting that a belief in objective values is built into ordinary moral thought and language, but holding that this ingrained belief is false. (Mackie 1977: 48–49)[1]

But what does Mackie mean by "objective values"? Is he referring to values that are objective in the sense that their value is independent of subjective valuings and desires? Does he mean objective as in scientific objectivity? As one can tell from the above passages, Mackie understands objective values as "objectively prescriptive"; they are a class of values that somehow possess an "intrinsic prescriptivity." This means that if a person would become aware of an objective value, then he or she would feel compelled to act in regard to that value, for instance by preserving it or protecting it. Thus when Mackie denies the existence of objective values, what he really takes issue with is the uniquely *motivating* aspect of objective values. In contemporary metaethics, *internalism* is a view about the relationship between moral judgments and motivation. Internalists say that there is a necessary relation between thinking that one morally ought to do something (making a moral judgment) and being moved to do it (Darwall 1998: 53).[2] For a moral internalist, morality necessarily carries with it a motivational component. Though internalism is not a term Mackie himself uses, it is clear that he characterizes judgments about objective values as an internalist would. By Mackie's understanding of objective values, in order for something to count as an objective value, when someone makes a judgment involving objective values, the judgment must be necessarily motivating. Therefore, when he says there are no objective values, what he means is that there are no values that are intrinsically motivating. Internalism about morality is a key premise, then, in Mackie's case for error theory.

Mackie lends further supports to his case by an "argument from relativity." He reasons that if there were objective values—values that intrinsically prescribe and motivate people to act—then why would we observe so much "variation in moral codes from one society to another and from one period to another"? (Mackie 1977: 36). He concedes that the variation and disagreement in moral codes does not provide *direct* evidence against objective values, but only *in*direct evidence. If people were really apprehending objective truths and values, then one should not expect to find such a high degree of variation and disagreement in ethics. For Mackie, it is more plausible to suppose that the variations in moral codes are the result of different ways of life rather than the result of dim perceptions of objective values. In short, the explanation for why there is so much relativity, in Mackie's view, is that there are no objective values. And since commonsense morality is committed to there being objective values, commonsense morality is in error.

Mackie believes that even more support for his error theory derives from "the argument from queerness." He says that the argument

> has two parts, one metaphysical, the other epistemological. If there were objective values then they would be entities or qualities or relations of a very strange sort, utterly different from anything else in the universe. Correspondingly, if

> we were aware of them, it would have to be by some special faculty of moral perception or intuition, utterly different from our ordinary ways of knowing everything else. (Mackie 1977: 38)

The reason why objective values would be so strange is because they would be "intrinsically action-guiding and motivating" (Mackie 1977: 49). Mackie thinks the way we could come to know these values is mysterious.

Another way Mackie brings out the queerness of objective values is to ask how objective values are linked or connected with the fabric of the world. Someone who believes in objective values must explain how these alleged entities are linked to natural features of reality. Mackie asks:

> What is the connection between the natural fact that an action is a piece of deliberate cruelty—say, causing pain just for fun—and the moral fact that it is wrong? It cannot be an entailment, a logical or semantic necessity. Yet it is not merely that the two features occur together. The wrongness must somehow be "consequential" or "supervenient." (Mackie 1977: 41)

For Mackie, the connection between objective values and natural features of the world is a mystery, as yet unexplained. In his estimation, the easiest way out of the difficulty is to reject the commonsense assumption that there truly are objective values that form the core and backbone of ethics.

Having mounted his case for error theory and against objective values, Mackie admits that he needs to provide an explanation of why the belief in objective values has persisted so long in popular opinion and ethical theory if there really are no objective values. He sketches several historical processes that give rise to what he calls "patterns of objectification." As the "apparent objectivity of moral value is a widespread phenomenon," he says it "has more than one source" (Mackie 1977: 46). He thus describes multiple patterns of objectification. An account of why people believe in objective values is a complicated picture; the following are some of the main elements Mackie uses to paint this picture. The erroneous belief in objective values is partially the result of people projecting their feelings and attitudes; the belief is also partially due to the fact that belief in objective values satisfies social needs and demands that arise in social communities; these social needs and demands exert pressures that over time become internalized in individuals. A desire for moral judgments to be authoritative is also part of the picture. Additionally, the traditional equating of right and wrong with divine commands tends to objectify values, and the persistence of the belief that ethics is a system of law, even after belief in a divine legislator has faded, will continue to cause people to believe in objective values. Disparate factors like these, says Mackie, have been responsible for the "objectification of moral values" (Mackie 1977: 42–46).

Mackie does not feel the need to specify the *exact* route that has led to the objectification of moral values; he is satisfied with his speculative account. This is partly because there is no *one* particular path to the objectification of value, but also because having already offered several arguments against objective values, at this point he only needs to show that some of his speculative patterns of objectification are possible. He finds an abundant supply of possible causes for numerous patterns of objectification; and these patterns can easily account for the persistent (false) belief in objective values. Before summing up Mackie's case for error theory, however, I must mention two more aspects of his metaethical approach. First, he frames the issue about the ultimate nature of ethics in ontological terms. His metaethical view "says that there do not exist entities or relations of a certain kind, objective values or requirements, which many people have believed to exist" (Mackie 1977: 17). Error theory makes the ontological claim that objective values do not exist, notwithstanding ordinary moral thought and language. Second, although Mackie is mounting a case against objective values, he has not argued that it is meaningless to talk about objective values; it is, rather, an error. We speak falsely when we refer to objective values.

> The assertion that there are objective values or intrinsically prescriptive entities or features of some kind, which ordinary moral judgements presuppose, is, I hold, not meaningless but false. (Mackie 1977: 39)

Cognitivism, another term from contemporary metaethics that Mackie did not use, is a term that applies here. Cognitivism is the view that moral judgments can be true or false, and so moral statements and value statements are the kinds of statements that can be true or false; in other words, they take truth values.[3] Mackie's claim is that moral judgments that presuppose the existence of objective values are always false; thus, the truth-value of objective value statements is *false*.[4] Mackie does not want to make the blanket claim that value statements in general cannot be either true of false. He explains why.

> For there are certain kinds of value statements which undoubtedly can be true or false, even if, in the sense I intend, there are no objective values. (Mackie 1977: 25)

The main ideas of error theory as developed by Mackie can now be summarized as follows. Commonsense morality presupposes the existence of objective values. But there are no objective values: there are no values that are intrinsically motivating. Mackie's argument from relativity says there is so much relativity and disagreement in ethics because there are no objective values. His argument from queerness says that objective values, if they existed, would be strange entities. How could we come to know them, and how would they connect up to natural features of the world? Patterns of objectification explain why people continue to believe in objective values when none truly exist. Mackie's error theory is an ontological thesis that incorporates motive internalism, which is a view about moral psychology, and cognitivism, a view about moral judgments and moral discourse.

2.2 ERROR THEORY AND EVOLUTION

In *Ethics: Inventing Right and Wrong* (1977) Mackie very briefly remarks on how he sees evolution fitting in with his view of ethics. He describes that moral sentiments—presumably those feelings he mentioned in his story about patterns of objectification—will most likely become widespread in the human species through the process of natural selection (Mackie 1977: 113). Mackie works with evolution and ethics in several articles as well (Mackie 1978, 1982a, 1982b). In these, for the most part he engages with normative ethics and evolution, matters such as competition, cooperation, egoism, and altruism. With regard to metaethics and evolution, in one of his papers Mackie argues that nonmoral retributive emotions can easily be imagined to have biologically evolved.[5] Mackie thinks that showing that nonmoral retributive emotions have their origin in evolutionary and subjective background elements, further supports his view about the nature of ethics, and casts doubt on those who view ethics as objective (Mackie 1982b: 219). When human beings living in social groups experience the uniquely social needs and demands that arise in social communities and those demands become internalized, strengthened and solidified, then nonmoral retributive emotions can be transformed into moral retributive emotions. Mackie thus goes some ways in showing how an evolutionary framework lends partial support to his hypothesis about patterns of objectification that have been partly fashioned out of feelings and sentiments.

The second generation error theorists, Michael Ruse, author of *Taking Darwin Seriously* (1998a) and Richard Joyce, author of *The Myth of Morality* (2001) and *The Evolution of Morality* (2006), more explicitly discuss error theory from the perspective of evolution. They both have argued that the fact of human evolution undermines the facts of morality and shows that morality is illusory and nothing but a systematic deception. They argue that error theory gains support when coupled with an evolutionary account of human origins. This is supposedly because a theory of evolutionary origins provides us with an explanation of why we are prone to fall into the error of believing in objective values and moral facts, namely, it is a helpful illusion to believe in such things. Ruse and Joyce both work with the question "if evolution is true, what does it imply about error theory?" And both argue that when we take evolution seriously, we will gain more support for error theory.

2.2a Ruse on Error Theory and Evolution

Ruse has published widely on the topic of evolutionary ethics and what evolution means for the foundations of ethics, and thus has been called "the most celebrated philosopher in the world for his untiring effort to join biology and ethics" (Rolston 1995: 90). Evolutionary ethics, as Ruse sees it, is "the project which argues that for a full understanding of the nature and grounds of morality one must turn to the process and theories of the evolutionist" (Ruse 1993: 133). Ruse is emphatic that the insights of biology can throw light on our understanding of ethics. To mark the distinction between traditional theories of evolutionary ethics that trace back into the nineteenth century, and the newer developments in the field, Ruse uses the label "evolutionary ethics" to refer to the traditional approaches and dubs his newer account "Darwinian ethics." Although traditional theories of evolutionary ethics were often wrongheaded, Ruse believes that a "Darwinian ethic" that goes back to the basics of Darwin's theory has much to offer those interested in the general nature and origins of ethics. Ruse notes that Darwin himself "was quite adamant that conventional morality can be given a selectionist backing" (Ruse 1989: 230). An example of how early evolutionary ethics was misguided, says Ruse, was in its supposition that evolution leads to higher and higher forms of life and culminates with human beings as the pinnacle of the natural world. The assumption that the process of evolution is directed does not have a proper place in the purely biological facts of evolution. Although "the traditional evolutionary ethicist argues that the process of evolution is not meaningless," "to the [Darwinian] ethicist, whose initial premise is the nondirectedness of evolution," the meaninglessness is indisputable (Ruse 1993: 139, 158). Darwinian ethics will accept the fact that people commonly think and feel that ethics is objective. There is no question for Ruse that an important part of the "experience of morality" is that it has "an aura of objectivity about it" (Ruse 1993: 153). But we cannot take that aura of objectivity at face value, he says, for we now have a scientific explanation of why human beings act morally or have a moral sense at all. Darwinian ethics says that moral behavior and moral sentiments owe their origins to the process of natural selection. A Darwinian perspective will view moral behavior as simply an adaptation: "We are moral because our genes, as fashioned by natural selection, fill us full of thoughts about being moral" (Ruse 1993: 148). Ruse's position is that a scientific explanation of human morality is now in our grasp and is wholly sufficient in accounting for the general nature and origins of ethics.

> The scientific claims are as simple as this. We now know that despite an evolutionary process, centering on a struggle for existence, organisms are not necessarily perpetually at conflict with weapons of attack and defense. In particular, cooperation can be a good biological strategy. We know also that humans are organisms which have pre-eminently taken this route of cooperation and working together. Further, there is good reason to think that a major way in which humans cooperate together is by having an ethical sense. Humans believe that they *should* work together, and so—with obvious qualifications—they do so. (Ruse 1993: 144)

Although it is commonly assumed that ethics is objective, Ruse believes that "if the empirical thesis . . . is right, there is no objective ethics. Nor can one readily see how the objectivist might patch up the situation, making his/her position compatible with evolutionism" (Ruse 1998a: 264). The objectivity of ethics, which is presupposed by ordinary moral thought, is an error, an illusion.

> We may have choice about whether to do right or wrong, but we have no choice about right and wrong themselves. If morality did not have this air of externality or objectivity, it would not be morality and (from a biological perspective) would fail to do what it is intended to do. . . . In a sense, therefore, morality is a collective illusion foisted upon us by our genes. (Ruse 1998a: 253)

This Darwinian account of the nature and origins of ethics, says Ruse, has undermined the foundations of ethics. The admission of the evolutionary origins of human beings compels us to concede that there are no foundations for ethics. Error theory says that there is an error in thinking that ethics has an objective component, that there are objective values. Mackie's notion of "patterns of objectification" is meant to explain *why* the masses and ethical theorists alike have continued to commit this error. Ruse, with a move similar to Mackie's, believes he can explain the belief and feeling that ethics is objective, as simply a product of our evolutionary past. Like Mackie, Ruse intends to avoid making ontological assumptions beyond those of evolutionary science.

Whereas Mackie calls the belief in objective values an error, Ruse calls it a collective illusion. And just as Mackie himself sought to bolster his notion of "patterns of objectification" with an evolutionary account, Ruse, too, pursues this angle. Ruse sees an even stronger role for biology to play in the development of "patterns of objectification," that is, in the explanation of why people persist in believing in objective values when ontologically there are none. Ruse puts the point in terms of genes, which he determines are the source of the collective illusion that morality is objective. In a nutshell, Ruse agrees with Mackie's assessment that commonsense morality presupposes the existence of objective values. Ruse further agrees with Mackie's ontological claim that there are no objective values. Ruse does not emphasize either the argument from relativity or the argument from queerness (although he does agree with Mackie that objective values would be ontologically queer).

For error theorists, the surface appearance of objectivity is an illusion; it is an error to take the objective feel of morality at face value. Ruse says Darwinism reveals why we commonly commit this error and are taken in by this illusion.

> Darwinism shows that, although morality may be all a question of feelings or sentiments, we humans project it into a prescriptively binding, supposedly objective status. We "objectify" morality. And because—and only because—we do this, morality functions as an efficient social facilitating mechanism. One who saw all of this, and who was (at the time of his death) relating it to evolutionary biology, was the late John Mackie. (Ruse 1998a: 277)

2.2b Joyce on Error Theory and Evolution

Joyce, a more recent proponent of error theory in ethics, has published a book-length defense of error theory (2001). His strategy for defending error theory differs from those we have seen so far. Mackie's error theory focuses on objective values and their intrinsic motivational capacity, which Mackie says is commonly believed to be inherently packed into them. We called that aspect of his theory motive internalism. Ruse's error theory focuses on the objective aura surrounding ordinary morality, the feeling and belief that morality is objective. Joyce's strategy for defending error theory focuses on reasons. While Mackie's case for error theory relies on *motive internalism*, Joyce's case for error theory relies on *reasons internalism*.[6]

Ordinary morality, says Joyce, presupposes that there are objective reasons why people should act in certain ways. He believes that, according to common-sense morality,

> when we say that a person *morally* ought to act in a certain manner, we imply something about what she would have reason to do regardless of her desires and interests. (Joyce 2001: 134)

According to Joyce, in ordinary moral discourse when we say that an agent morally ought to do something, we presuppose that there are genuinely unconditional, categorical, moral imperatives and the agent is bound to their moral authority, regardless of his or her desires. Ordinary moral discourse presupposes that there are objective reasons that make morally obligatory actions inescapable: the agent cannot evade the obligation by claiming not to have a desire to perform the action. The very nature of morality is categorically binding. The upshot of Joyce's position is that, while morality carries this authority and needs to have this authority in order to work, the claim that there are indeed objective reasons that underwrite the categorically binding nature of morality is ultimately indefensible (Joyce 2006: 198–99). Common-sense morality presupposes that there are objective reasons for an agent to act morally; but common-sense morality is premised upon an error, because

> after careful investigation we have found no defensible grounds for thinking that such reasons exist. (Joyce 2001: 134)

According to Joyce, we are in error to think that there are genuinely unconditional, categorical, moral imperatives.

Error theory says that the common understanding of morality is in error, but error theory does not supply us with an explanation of why we are prone to fall into this error, says Joyce (Joyce 2000: 725; 2001: 135). But Joyce says that the fact of evolution can help fill this explanatory gap because evolution can provide an explanation of why we are led into error. If we interpret morality as an evolved trait, he says, then we are easily led to the conclusion that morality—with its assumption about unconditional categorical imperatives—is a systematic error. In short, when error theory is coupled with an evolutionary account of human origins, error theory becomes more plausible.

But even if categorical imperatives are ultimately ungrounded, they are still evolutionarily useful (Joyce 2000: 714). Joyce employs the concept of a *useful fiction* (Joyce 2001, chapters 7 and 8). He accepts that morality is a useful fiction; he does not believe that if we accept error theory we are then rationally obliged to abolish our commitment to morality. Moral fictionalism works fine for us, he says. Practical morality with its assumption about objective reasons is serviceable, though, at bottom, untrue. Overall, in the light of the fact of human evolution, for Joyce, error theory is the most plausible metaethical account.

2.3 MODELS OF EVOLUTION AND ERROR THEORY

2.3a Darwin and Error Theory

Darwin's model of evolution by natural selection is a unilevel theory; he holds that natural selection selects for individual organisms, not for species, communities, or ecosystems. When it comes to human ethics, then, one would think that Darwin would characterize ethics as an adaptation that exists to benefit individual human beings. He does not do this, however. Ethics is an area where Darwin employs a group selectionist model (Sober & Wilson 1998: 4). Perhaps we can get some idea of what Darwin's model of evolution implies about error theory by considering what Darwin himself said about evolution and ethics. In chapter 4 of *The Descent of Man* (1871), Darwin discusses a wide variety of ethical issues, exclusively, he says, from the perspective of natural history (Darwin 1871: 41). He discusses various qualities of human beings that are intimately connected with ethics; he calls these qualities collectively "man's moral constitution" (Darwin 1871: 71). These human characteristics include: a moral sense, a conscience, social instincts, parental and filial affections, virtues, sympathy, and other emotions like love, remorse, regret, and shame. And Darwin discusses ethical issues such as the relationship of pleasure and pain to ethics, the place of reasoning in ethics, motives for human moral actions, and the "foundation of morality" (Darwin 1871: 64, 71). He raises questions about these issues, such as "where do sympathetic feelings come from?" (Darwin 1871: 52–53); and "are these kinds of characteristics in human beings the result of habit or natural selection?" (Darwin 1871: 50–51). In numerous places he concludes that, although habit can strengthen various human qualities, natural selection did likely have a role in how human beings came to possess many of these characteristics in the first place (Darwin 1871: 49–50, 52). Darwin certainly agrees with Mackie that there is such a thing as commonsense morality. In fact, Darwin takes it to be highly probable that

> any animal whatever, endowed with wellmarked social instincts, the parental and filial affections being here included, would inevitably acquire a moral sense or conscience, as soon as its intellectual powers had become as well, or nearly as well developed as in man. (Darwin 1871: 42)

The very concept of "a moral sense" that Darwin uses seems to be shorthand for "commonsense morality." Darwin, when discussing ethics, then, takes as his subject matter not only morality as it exists in nineteenth-century England, but morality as it is for early hominids, and any social animal with developed social instincts and intellectual powers. Darwin has provided an evolutionary naturalist's "view of the origin and nature of the moral sense, which tells us what we ought to do" (Darwin 1871: 61). The following are aspects of commonsense morality that Darwin discusses: conscience, sociability, duty, habit, rules/standards of conduct, love and sympathy for others, defending oneself, approval and disapproval of others, services for others, sacrifice for others, and defending others.

A key piece of how Mackie supports error theory is his finding that commonsense morality presupposes the existence of objective values, values that are intrinsically prescriptive. In Darwin's observations about ethics, does he too observe that as a prominent feature of commonsense morality? More specifically, does Darwin's evolutionary perspective on ethics point toward motive internalism as the accurate way to describe the psychological dynamic between moral values, motivation, and action? Motive internalists hold that there is a necessary relation between thinking that we ought to do something and being moved to do it. When Darwin writes of being impelled to particular lines of conduct, he seems to be holding an internalist characterization of ethics (Darwin 1871: 53). Mackie is careful to make a distinction between nonmoral and moral retributive emotions, with the latter described as relying on a more developed social atmosphere. Darwin, too, emphasizes that being "impelled" "to certain lines of conduct" is not a motive innately implanted in a human being by natural selection, but rather a motive that can likely develop once other aspects of human nature such as feelings, powers of reasoning, and a social environment have been put into place. When it comes to ethics, Darwin does acknowledge that human beings can be impelled by an instinctive motive to perform self-sacrificial acts to aid others (Darwin 1871: 54). Darwin does not use the phrase "objective values," and it may be going a bit far to say that he thinks commonsense morality presupposes an ontological claim that "objective values exist." But the tug of morality, the "being impelled" of morality that has deep roots—deeper than mere habit can explain—is something that appears in Darwin's description of ethics, which is written from the perspective of natural history.

If Darwin's model of evolution does fit with Mackie's characterization of commonsense morality and objective values, then Mackie's arguments from relativity and queerness retain their force. For when we philosophically reflect on the objective tug of morality, we need to ask why, if there are objective values that motivate human beings to act morally, there is so much relativity in ethics? And, if there really were objective values, how do we know about them, and how do they connect with the natural features of the world?

We will recall that Mackie, when defending his error theory, supplemented a small portion of his error theory with evolutionary considerations. His notion of "patterns of objectification" as a proposed explanation for why we fall into an error about morality, gains support, says Mackie, by viewing part of the story of the objectification of values as intertwined with the evolutionary development of emotions and moral sentiments. Human beings, equipped with a suite of emotions and immersed in a social dynamic of community living, will come to objectify values. Ruse, who says he is relying on an explicitly Darwinian model of evolution and adopts the phrase "Darwinian ethics" to underscore that fact, agrees that when we objectify morality and presume that it has some kind of standing independent of biological adaptation, we are bewitched by a collective illusion.

To see what Darwin's model of evolution implies about Joyce's error theory let us look at some other passages from Darwin's *The Descent of Man.* Darwin notes that some people believe that in order for an action to be legitimately regarded as a moral action, it cannot be "performed impulsively" but must be "done deliberately after a victory over opposing desires, or when prompted by some exalted motive" (Darwin 1871: 55). Darwin calls this the view of "formal morality." Since he has described actions in which "moral beings" are "impelled by an instinctive motive to help others," he thinks his view differs from a formalist view of moral agency. Although Darwin agrees that someone who struggles against fear, for example, and yet performs a heroic action, will deserve more credit, he does not think that struggling against opposing desires should be the criterion of a moral action. He gives several reasons for this, one of which points to the common occurrence that if we perform a certain action very often we will not need to deliberate or hesitate with that type of action any longer, it will become second nature to us. If it is a good action, should we refrain from calling it moral because we can now perform it effortlessly due to repeated performance? Darwin thinks not.

The position Darwin takes against "formal morality" bears on Joyce's strategy for defending error theory. Joyce's error theory, which is premised upon reasons internalism, views commonsense morality as presupposing that a moral agent is bound to the authority of categorical moral imperatives, regardless of his or her desires. Because *genuine* categorical imperatives are indefensible, argues Joyce, commonsense morality is fraught with error. It seems that Joyce characterizes commonsense morality as formal morality, a characterization Darwin finds questionable. Rather than characterize moral actions as always opposed to desires, Darwin suggests that an action can be considered moral if it is performed by a being "who is capable of comparing his past and future actions or motives" (Darwin 1871: 55). An action performed by such a being is considered a moral action "whether performed deliberately, after a struggle with opposing motives, or impulsively through instinct, or from the effect of slowly gained habit" (Darwin 1871: 56).

If Darwin, based on his understanding of ethics and social animals, does not acknowledge that commonsense morality presupposes that objective, categorical reasons can be exacted for all moral actions, then he disallows Joyce's first premise. Perhaps natural history is not the place to look for support for reasons internalism. A Darwinian view of evolution will acknowledge that human beings feel a tug of morality, and are sometimes impelled by morality. Even if it is not felt in every moral action, the pull of morality does provide a background context in the sense that it is one of the basic components of commonsense morality.

2.3b Dawkins' Model of Evolution, and Error Theory

Let us now think of evolution in terms of Dawkins' model and consider what it would imply about error theory. Like Darwin, Dawkins has a unilevel model of natural selection, but Dawkins focuses on the level of the gene;

he understands evolution in terms of genic selection. When trying to explain the evolution of some characteristic, Dawkins says we need to think in terms of gene frequencies. We should understand an organism as a vehicle and an organism's characteristics as phenotypic strategies for perpetuating genes (small pieces of DNA). Human beings, as individual organisms, are thus vehicles; and so we should think of human behavior as the behavior of vehicles. The theory of evolution, along with the auxiliary hypotheses of kin selection and reciprocal altruism, provides a naturalistic explanation of why humans exhibit the behaviors and possess the characteristics they do, including ethical behaviors and what Darwin called "man's moral constitution" (Darwin 1871: 71).

According to Dawkins, though, genes have only an *in*direct effect on human behavior, a statistical influence. Genes provide behavior rules for their vehicles to (statistically) follow. When I act I don't have time to do calculations concerning what will be best for my genes. And anyway, as a vehicle, I am not conscious of these rules.

An example of how genes program their vehicles to follow rules of behavior that protect the interest of the vehicle's genes is the behavior known as "distraction display." Many ground-nesting birds do this by limping away from their nest when a predator approaches. The parent bird lures the predator away from the nest where the chicks are (Dawkins 1989: 6). The parent bird's "distraction display" behavior may have brought a risk to itself, but from a genic selectionist perspective, if the bird's genes manage to make their way into more vehicles (the parent bird's offspring), it is worth the risk.

Now with regard to metaethics, in order to get to Mackie's first step toward error theory, we need to ask if Dawkins' model of evolution throws any light on Mackie's contention that commonsense morality presupposes the existence of objective values. Dawkins occasionally writes about commonsense morality (2001, 2005, 2006a). In describing how the behavior of organisms is related to genes, Dawkins' genic selection could, it seems, be used in support of error theory's notion that human beings maintain beliefs about objective values, not because they are true, but due to other causes.[7] Mackie speculates about these other causes with his "patterns of objectification."

Human beings have the belief that there are objective values, but that belief is false, says Mackie. Does possession of that belief—even though it is false—lead to beneficial consequences? Dawkins holds that the ultimate value of genes is to replicate more genes (Dawkins 2001: 2, 12). Would such a goal be advanced by programming vehicles to be disposed to believe in objective values? It does seem that such a strategy would make sense on genic selectionist grounds.

Consider the vignette at the beginning of this chapter; it is a case in which genes have programmed their vehicles to follow a particular rule of behavior that protects the interest of the vehicle's genes. Also, though, the rule causes the vehicle to commit an error. Still, in the long run the error is helpful to the vehicle's genes. Dawkins says there is a rule built into bird parents that says, "Be nice to any small bird sitting in the nest that you built" (Dawkins 1989: 102). Meanwhile, "cuckoos," known as a kind of brood-parasite, will "lay their eggs in somebody else's nest" (Dawkins 1989: 102). Foster-parent birds are then tricked into caring for the cuckoo chicks. The foster-parent birds are committing an error by extending their efforts and energies into caring for cuckoo chicks, for the cuckoos do not possess the foster-parent bird's genes. Although the foster-parent birds will commit this error when they follow the rule "Be nice to any small bird sitting in the nest that you built," on the whole, it is still a beneficial rule for the foster parent's genes to program its vehicles to follow. In the long run, the foster-parent bird's genes still statistically compete well enough in the struggle for genic survival. When Ruse makes his case for error theory based on evolutionary considerations he seems to have genetically implanted rules like these in mind. He says morality is an illusion fobbed off on us by our genes.[8]

The analogy between bird behavior and human behavior is not watertight, of course. Dawkins is clear that genes have only an indirect influence on human behavior. Because of the influence of culture on human behavior, the human species is quite unique, thinks Dawkins (Dawkins 1989: 189). Dawkins even goes far as to say that "man's way of life is largely determined by culture rather than by genes" (Dawkins 1989: 164). If Dawkins' model of evolution when applied to human behavior includes the important qualification that the human way of life "is largely determined by culture rather than by genes," then where do we stand with respect to error theory?

The first important issue in looking at error theory through the lens of Dawkins' model of evolution is that Dawkins does not explicitly take a position on whether commonsense morality presupposes the existence of objective values. So we cannot say that Dawkins' model provides direct support for the first important premise of Mackie's error theory. There isn't even support for Ruse's claim that ethics has an objective aura about it. And there is definitely no hint of support for Joyce's reasons-internalism defense of error theory. What we do find in Dawkins' model of evolution is more support for Mackie's explanation of why we *would* commit an error about the objectivity of values, if we did. Dawkins' notion that genetic rules constrain and direct behavior for the purposes of the genes that are at the helm helps to see how it could be the case that behavioral rules put in place by evolution could direct us into a systematic error, though an error that could be beneficial in some way. However, since human behavior, by Dawkins' estimation, is not wholly directed by biological rules, it could be the case that we commit the error about the objectivity of values more so because of the effect of culture. In the end, it seems that Mackie was correct when he insisted that there are multiple "patterns of objectification" and several historical processes that could account for why human beings fall into the error of believing in objective values (granted that they do). Genic selection is but one historical process; there are many other cultural processes that direct human belief and behavior. Darwin, too, displays this caution when he wonders whether a given human characteristic is a product of habit or natural selection. Ruse, who sees moral behavior as a biological adaptation, does not seem to allow for culture's role. Given Dawkins' claim about culture and human behavior, it seems that, by the model of genic selection, Ruse has actually overstated evolution's support for denying objectivity in ethics. Ruse says that for a full understanding of the nature and grounds of morality we must turn to the evolutionist because the insights of biology throw light on our understanding of ethics. But when we turn to Dawkins the evolutionist, Dawkins tells us that *culture* may have more influence on human behavior than biology. That does not seem to be what Ruse had in mind.

The mixed bag of support that Dawkins' model of evolution supplies for error theory is also evident with Joyce's defense of error theory. On the one hand, genic selection offers (1) a conceptual apparatus to explain why human beings would fall into the error of believing that morality is unconditional, and (2) a working model of how fictions (and errors) can be genetically useful. On the other hand, though, genic selection doesn't show that (3) commonsense morality presupposes a belief in objective reasons, and (4) it does not establish that human biological evolution, not cultural evolution, throws commonsense morality into this error, as Joyce says it does.

2.3c Gould's Model of Evolution, and Error Theory

Let us consider error theory from the perspective of Gould's model of evolution.

About Mackie's main case for establishing error theory, Gould's model of evolution seems neutral: evolutionary theory simply is not in the business of identifying the assumptions of commonsense morality. When we turn to Mackie's notion of patterns of objectification, the second phase of his defense of error theory, in which he offers an explanation of why human beings would believe in objective values while there are none, Gould's model of evolution may have some relevance.

Mackie suggests various mechanisms to account for patterns of objectification: people projecting their beliefs and attitudes; nonmoral retributive emotions acquired through natural selection solidifying into moral retributive emotions when culture has time to work with them; belief in objective values satisfies social needs and subsequently becomes internalized; the desire for an authoritative source for morality; the traditional connection of divine commands with right and wrong, etc. What would Gould make of such an account?

Gould has an abiding appreciation for contingencies and history, and for the contingencies *of* history, and he thinks that Darwin, too, was committed to viewing contingency as a central aspect of natural history (Gould 2002: 1333, 1336). Mackie's suggested patterns of objectification seem to be paradigm examples of how a particular phenomenon, in this case a purported widespread commonsense belief, may ultimately have nothing more behind it than a series of haphazard and contingent meanderings of cultural twists superimposed upon a few chance biological turns. Gould is usually suspicious of how evolutionary theorizing is applied to human beings, because human behavior is so flexible and unpredictable, while evolutionary theorists often attempt to describe *the* explanation of a particular human behavior. Only versions of "crude biological determinism," says Gould, will postulate "a direct genetic basis for our most fundamental traits" (Gould 1977: 238). In this case, though, Mackie's speculations about "patterns of objectification" with an evolutionary backing do not seem to give a fixed picture of how human beings presumably came to hold this particular belief about objective values. Instead, Mackie is simply asking us to imagine that a combination of particular biological predispositions and the molding forces of sociocultural context account for a persistent belief in objective values, although no such values really exist.

When it comes to Ruse's claims that moral objectivity is an illusion and that morality has been fobbed off on us by our genes, Gould's evolutionary model would perhaps say that Ruse is being too deterministic in his application of the evolutionary theory to human thought and behavior. As noted above (§2.3b), Ruse does not seem to leave much room for culture as the source of the aura of objectivity. It may indeed be the case that commonsense morality has an aura of objectivity about it, as Ruse says, but there is nothing in Gould's model of evolution that implies that this aura is simply a collective illusion, or that its being illusory is the product of our genes. According to Gould, the human brain as it became larger also became overqualified for the role it was playing for early humans; consequently, human beings now have the intellectual capacity to be conscious of their actions and their lives. Human beings now feel inclined to create meaningful narratives about themselves that aid them in coping with their biological existence. "Patterns of objectification" would seem, then, to make much sense from Gould's perspective, especially in the manner described by Mackie, who leaves much more ample room for the cultural, not necessarily biological, causes. Let us now consider error theory as developed by Joyce. Does Gould's model of evolution reveal that commonsense morality presupposes that there are objective reasons that are in tension with our desires? Not directly; for again, evolutionary theory is not in the business of identifying the assumptions of commonsense morality. Occasionally, in his essays, though, Gould will discuss ethical concepts, principles, theories, and traditions. From his perspective, a morality consisting of absolute and unconditional categorical imperatives does not fit well in a "complex and sloppy" world like ours (Gould 1993: 50). Joyce characterizes morality as an unconditional constraint that supplies objective reasons, which prescribe the morally right thing to do irrespective of one's desires. Joyce then says that, since no such objective reasons exist, therefore morality is premised upon an error. Here we might think that Gould would agree with Joyce's error theory because Gould also views unconditional categorical imperatives as implausible (perhaps queer) demands in a naturalistic world.

But Gould, after he notes that a morality of absolute and unconditional categorical imperatives seems implausible, then describes *other* types of moral principles. Gould proposes that other moral principles, which are premised upon "desire, negotiation, and reciprocity," have appeared in many cultures in many different eras, and have shown their wisdom (Gould 1993: 50). If so, then Gould would not agree that commonsense morality in its entirety is premised upon an error, but only that a particular narrow conception of morality is premised upon an error, namely, the view that morality consists of rigorous, unbending demands that are always poised antagonistically against one's desires.

This point ties back to the earlier discussion about Darwin's position against "formal morality" (§2.3a). There, I said that Darwin would disagree with Joyce's characterization of commonsense morality. On this issue about morality, Darwin and Gould are in accord. Both of them find that morality when characterized as a set of formal constraints that are opposed to desires is not a conception of morality in accord with a naturalistic perspective. By Joyce's interpretation of the matter, the fact that unconditional categorical imperatives are indefensible shows that commonsense morality rests on an error; but for Gould (and Darwin), the fact that unconditional categorical imperatives are indefensible does not challenge commonsense morality, for commonsense morality need not be wedded to such a conception in the first place. In general, the ethical outlook Gould favors seems inconsistent with the notion that human beings are systematically mistaken in their beliefs about commonsense morality.

2.3d Haught's Model of Evolution, and Error Theory

While Darwin, Dawkins, and Gould have naturalistic models of evolution, Haught defends a theistic model. So before turning to what Haught's model implies about error theory, we must first clarify what Haught's model implies about naturalistic accounts of ethics generally. Haught's strategy for integrating evolution and theism is to accept methodological naturalism while rejecting metaphysical naturalism. He agrees to what the sciences say about the evolution both of life and the entire cosmos. In looking at the big picture of cosmic evolution, Haught employs a narrative cosmological principle: he sees cosmic evolution in terms of a story, one that involves a God of love. As a God of love, the creator allows creation to develop at its own pace and in its own way, including all manner of contingent twists and turns and an inordinate number of evolutionary dead-ends.

Related to his narrative cosmological principle, Haught also uses an *aesthetic* cosmological principle (Haught 2000: 128). As Haught sees it, the universe—seen in terms of geological time—is involved in the ongoing creation of beauty: as there is more complexity in the pageant of natural history, there is more nuanced beauty. He understands beauty as "a delicate synthesis of unity and complexity, stability and motion, form and dynamics" (Haught 2000: 131).

Human beings are part of this unfolding cosmological story. Thinking about ethics from a perspective that views all reality as an unfolding and evolving cosmos requires us to view human beings and human behavior as products of this evolutionary unfolding. In *God After Darwin* (2000), Haught has a chapter on religion, ethics, and evolution, and in both *Is Nature Enough?* (2006) and *Making Sense of Evolution* (2010) he has a chapter on morality. Although most of natural history saw the intensification of beauty without human involvement, now that human beings are on the scene—and have been for tens of thousands of years—human conduct can contribute to the intensification (or deterioration) of beauty in the cosmos. The aspect of evolution that Haught most emphasizes is that the evolutionary process is an unfinished and undirected story; as such, human action can make a difference in how the story unfolds and ultimately turns out.

In Haught's opinion, ethics cannot be fully naturalized: naturalists cannot construct ethics out of purely naturalistic terms (Haught 2006: 148). This is an issue that reaches into one of Haught's main arguments against naturalism. He introduces the argument in *Deeper than Darwin* (Haught 2003: 97–99), but more fully develops it in *Is Nature Enough?* (2006). The argument has two parts; one concerns the abilities of the human mind and the other concerns ethics.

First, as an evolutionist about the human mind, Haught insists that our mental capacities have come from the earth, through evolution (Haught 2006: 95). Nevertheless, there is a problem encountered when naturalism attempts to account for the subjectivity and critical intelligence of the human mind. For Haught, the human mind spontaneously performs different acts of cognition and it "*cannot help* passing through the three distinct but complementary acts: experience, understanding and judgment" (Haught 2006: 32–33). Experience, understanding, and judgment are the main elements of subjectivity. Collectively, Haught calls these features of the human mind *critical intelligence*. Now if we look upon critical intelligence, which is really just a natural ability and motivation to seek the truth, as "an incidental, unplanned byproduct of evolution" we will have "failed to justify the *trust*" we put in this intellectual capacity (Haught 2006: 112). Haught believes that naturalists who view critical intelligence as an accident of evolution undermine the confidence we place in the spontaneous workings of critical intelligence (Haught 2006: 113). Since we *do* spontaneously put trust in our own minds and we *are* confident in our intellectual capacities, the human mind's experience of subjectivity is something naturalism cannot account for (Haught 2000: 88). In a much stronger sense, naturalism is actually inconsistent with the trust we place in the imperatives of our minds (Haught 2006: 36).[9] Haught regards this as a self-contradiction in evolutionary naturalism (Haught 2006: 107); and he takes it to be such a strong argument against naturalism that it "spells the end of naturalism as a plausible creed" (Haught 2006: 88). I think he is a bit too overconfident about the strength of this argument, but I won't take the time here to evaluate the argument because we are here more concerned with his argument about ethics, not the nature of the human mind.[10] Theistic evolution has an advantage over naturalistic evolution, Haught says, because when faced with explaining how human critical intelligence can have such confidence in itself, a theist can say that the human mind has been grasped by a transcendent truth. The reason that we as intelligible subjects spontaneously put so much trust in the desire to know is because of our mind's native anticipation of a transcending fullness of truth that has *already* grasped hold of us (Haught 2006: 91).

But let us turn to see what this has to do with ethics. In your own mind right now, the three spontaneous acts of cognition (experience, understanding, and judgment) seem to be following "persistent and ineradicable imperatives at the foundation of your consciousness" (Haught 2006: 33). The persistent and unavoidable imperatives are: "be attentive," "be intelligent," and "be critical," respectively. But at work in your mind right now there is another cognitional act, one that has the most to do with ethics: it is the cognitional act of "decision," and *it* is driven by the imperative to "be responsible." Haught contends that in the natural imperative to *be responsible*, the human mind anticipates the future. And here comes the kicker, the stumbling block in trying to harmonize Haught's model of evolution with naturalistic ethics: a naturalistic ethics that tries to conjure the whole of ethics out of purely naturalistic materials without providing an explanation concerning the source of the native imperative to be responsible will inevitably come up short, contends Haught.

For Haught, most naturalists focus too much attention on genetic explanations of ethics and human behavior (Haught 2000: 132). He sees this as part of a "reading problem": there is a big difference between biology and biography, he says. While physical evolutionary explanations are necessary to understand critical intelligence, they are not sufficient to explain the trust we put in our own minds (Haught 2006: 90).

For Haught, the same holds true of ethics:

> The naturalist's own moral aspirations cannot be explained fully in naturalistic terms any more than can critical intelligence. (Haught 2006: 148)

If Haught's model of evolution is accurate, then what does it imply about error theory in ethics? Consider Mackie's error theory first. The pivotal first premise of Mackie's error theory is his claim that commonsense morality presupposes the existence of objective values. And Mackie understands objective values along the lines of motive internalism. Haught, with a model of human evolution that identifies in human beings a native imperative to be responsible—an unrelenting imperative due to the human mind being grasped by a transcendent truth—would seem to agree with Mackie about this. Mackie's second step in seeking to defend an error theory is to offer the arguments from relativity and queerness. If Haught concedes that there are indeed objective values of the kind Mackie has described, then the argument from relativity will demand that Haught explain how it could be the case that each person's human mind has a native imperative to be responsible and yet so much diversity in ethics would persist. Haught would most likely respond that the argument from relativity does not prove that beliefs about objective values are false. The same imperative is present in all human minds, he would say; it is only that local circumstances give rise to unique cultural histories, which in turn yield differing customs and ethical standards. (This is a response to the argument from relativity that Mackie himself considers but finds that it only partly counters the argument from relativity.)

Haught will then need to respond to Mackie's argument from queerness: if there were objective values, they would be queer entities in an otherwise naturalistic world. Again, Haught is likely to readily admit that objective values are indeed queer entities; but that will not lead him to suspect that they do not exist. Evolution is an ongoing process that has given rise to complexity, nuance, and beauty, and human nature has been crafted by this inventive evolutionary process; so under this model of evolution the presence of queer entities such as objective values that have a special capacity to motivate human beings into action is not altogether inconceivable. When Haught says that a naturalistic ethics is unable to conjure the whole of ethics out of purely naturalistic materials he is providing conceptual space for queer objective values. Since Haught would not view objective values as inconsistent with his model of evolution, he will not conclude that all statements and beliefs about objective values are false; hence, he will not need Mackie's "patterns of objectification." That aspect of Mackie's error theory tries to answer the question: If there are no objective values, then why would so many people believe in them? This issue won't come up for Haught, because he seems to believe that there are objective values (as defined by Mackie) and that people do in indeed believe in them, since he agrees with Mackie that commonsense morality presupposes their existence. Turning to Ruse's defense of error theory, it would seem that Haught's model of evolution does not support Ruse's claim that morality is a collective illusion foisted upon us by our genes, since Haught has a different explanation for the uniqueness of ethics (and the mind) in a natural world. For Haught, most naturalists focus too much attention on genetic explanations of ethics and human behavior; Haught's theistic model need not resort to viewing morality as an illusion because it has an objective feel to it and as such is difficult to accommodate into a naturalistic model of evolution. Haught believes that his theistic model of evolution accounts for why morality is an intransigent phenomenon for naturalism. Interestingly, Haught will agree with Ruse's claim that for a full understanding of ethics we must turn to the process and theories of evolution. Haught will hasten to add that only a theistic model of evolution, though, will truly yield a full understanding of ethics. Ruse will likely point out that a theistic model of evolution conflicts with contemporary evolutionary theory since contemporary evolutionary theory views evolution neither as directed nor meaningful. Haught's account of evolution does not view evolution as directed, though; he accepts that (cosmic) evolution is a contingent historical process unfolding in its own unique way. The cosmos is self-actualizing, and Haught contends that its meaningfulness is apparent when we are careful to avoid the reading problem (that naturalists suffer from), and we employ a narrative cosmological principle, and an aesthetic cosmological principle. For Haught, an evolutionary perspective does not undermine the foundations of ethics. Quite the reverse, in fact: the aura of objectivity surrounding ethics is a clue to a transcendent truth that falls through a metaphysical naturalist's net. In short, Haught's model of evolution will respond to Ruse that, yes, there is an aura of objectivity surrounding morality, but that is not evidence of an error at the heart of ethics.

In Joyce's defense of error theory he relies on reasons internalism, the view that ordinary moral discourse presupposes that there are objective reasons that make morally obligatory actions inescapable. Commonsense morality, says Joyce, presupposes that morality is about categorical imperatives. Haught's model of evolution, which incorporates his notion about a native imperative to be responsible, would seem to agree with Joyce on this. Joyce then argues that, since the claim that there are such objective reasons is, at the end of the day, indefensible, morality is premised upon an error. What does Haught's model of evolution imply about the prospects of identifying objective reasons that would make morally obligatory actions inescapable?

The first qualification Haught would put on Joyce's characterization of categorical imperatives is that Joyce has overstated the antagonistic relationship between categorical imperatives and desires.[11] I don't think Haught would agree that categorical imperatives, in order to be considered moral imperatives, must always be in conflict with desires. Haught is in agreement with the ancient observation that in human nature there is a natural yearning for the good (Haught 2000: 122). If human beings have a natural yearning for the good, then in some deep sense they desire the good. If so, then the moral thing to do is not wholly at odds with the desirable thing to do.

But, specifically about objective reasons and reasons internalism, if our starting point is Haught's model of evolution, which it is, at least in this section, then we should bear in mind his narrative and aesthetic cosmological principles. Under Haught's model of evolution, human conduct can contribute to the intensification (or deterioration) of beauty in the cosmos. And since the evolutionary process is an unfinished and undirected story, human action can make a difference in how the story unfolds and ultimately turns out. These two principles taken together would seem to provide the materials for filling in the objective reasons why human beings should engage in moral conduct. The kind of objection that Joyce makes against those who would defend objective moral reasons would, of course, also apply. Joyce would say that a given individual may have no reason to refrain from contributing to the deterioration of beauty in the cosmos. The idea here is that, even if the person were to accept the plausibility of the aesthetic cosmological principle, the principle does not provide *this particular individual* with a reason to refrain from diminishing the amount of beauty in the cosmos. Haught would likely respond that what makes a principle or a reason objective in the first place is precisely that it doesn't matter that a particular individual doesn't subjectively and consciously possess the reason. What matters is whether it is possible for the individual to gain access to the reason and the principle; if it is not possible for the agent to come to have the reason, then by the principle of ought implies can, the agent cannot justly be asked to abide by the reason. But if the reason *is* accessible to the agent, which in this case Haught says it is since it is predicated on his model of evolution, then it makes sense to say that the agent is bound by objective reasons to fulfill his or her responsibility and contribute to the intensification of beauty in the cosmos, not to its diminution. I don't think Joyce will find this line of argument convincing. But my concern in this section is with what Haught's model of evolution implies about error theory; so, for my purposes, I am more interested in what Haught's model of evolution implies about Joyce's error theory than in what Joyce's error theory implies about Haught's theory of evolution. Overall, I think it is safe to say that Haught's model of evolution does not lead us to suspect that commonsense morality is in error or is simply a useful fiction.

2.4 CONCLUSION

In this chapter I first explained the main elements of error theory in the terms in which Mackie originally set them out. I also looked at the error theories of Ruse and Joyce. I then described how Mackie views that evolutionary thinking lends partial support to his hypothesis about patterns of objectification. After Mackie, both Ruse and Joyce specifically worked with the question, if evolution is true, what does it imply about error theory? Ruse says that Darwinism reveals why we commonly commit the error of presuming that morality is objective and are taken in by this illusion. Joyce, espousing an error theory based on reasons internalism, says that evolution tells us why we fall into such an error: morality is an evolved trait and a useful fiction. Whereas Mackie claimed that in ethics "there are no objective values," Joyce says in ethics "there are no objective reasons."

In considering the implications of Darwin's model of evolution on error theory, I looked at Darwin's thoughts on ethics and evolution. Since Darwin acknowledges that ethics can impel human beings into action, Mackie's arguments premised on motive internalism are granted their starting point. Darwin takes a position against formal morality, though, so he seems to stand opposed to Joyce's version of error theory that relies on reasons internalism. When Dawkins writes on evolution he doesn't have much to say about commonsense morality, whether it carries an aura of objectivity or whether it presupposes objective values or objective reasons. But Dawkins' gene selectionism does seem to work with a theory that says ethics carries a kind of error and is only a fictional device that serves the interests of genes. However, since Dawkins claims that human behavior is largely determined by culture, the patterns of objectification that Mackie speaks of may have more to do with cultural than biological pathways. Dawkins' model of evolution seems to only provide a mixed bag of support for error theory.

Gould's model of evolution doesn't say whether commonsense morality presupposes objective values or objective reasons, but given Gould's position on moral principles in different cultures and eras, he does not seem to characterize ethics as error theorists require in order for their arguments to hit their mark. As with Dawkins, the patterns of objectification may have as much to do with culture as with biology, so an evolutionary perspective does not seem to provide special, additional support for error theory.

After introducing Haught's thinking about naturalism and ethics, I concluded that Haught would agree with Mackie that commonsense morality presupposes objective values. But I said that recognizing such a presupposition at the heart of ethics would not lead Haught to an error theory. Evolution most certainly has implications for ethics, but from the perspective of his model of evolution, this does not lead us to suspect that commonsense morality is in error or is simply a useful fiction. Before I bring this chapter to a close, I will make a few more points about evolution and error theory. In discussing error theory and evolution, the key starting point is, of course, with Mackie. Mackie's case for error theory begins with the claim that there are no objective values, that all moral values are subjective. The title Mackie gives to his main book in ethics (the book to which he frequently refers in his articles) is: *Ethics: Inventing Right and Wrong.* Since there are no objective values in the fabric of the world, human beings must invent ethics, right, and wrong, out of their subjective values. Invention, as opposed to discovery, captures the essence of ethics, and, as Mackie says, is central

> for someone who holds, as I do, that morality is a human product, that it is a system of thought and evaluation and control of conduct into which human feelings and desires and instincts and social interactions and reciprocal pressures enter, along with knowledge and beliefs of various sorts. (Mackie 1982a: 152)

Yet Mackie qualifies what he means with his claim that ethics is invented: "It does not follow," he says,

> that an individual is free to invent a moral system at will. If a morality is to perform the sort of function described in Chapter 5, it must be adopted socially by a group of people in their dealings with one another. Of course, there can be and are larger and smaller social circles. (Mackie 1977: 147)

Even though Mackie maintains that morality is a human product and something invented by human groups, not necessarily individuals, he also wants to claim that "human morality itself as the product of some kind of natural evolution" (Mackie 1978: 121). Mackie writes:

> It will be reasonable to ascribe to biological evolution those pre-moral tendencies to care for children and close relatives, to enjoy the company of fellow members of a small group, to display reciprocal altruism and both kindly and hostile retribution, which we share with many non-human animals, but to ascribe to cultural evolution the more specifically moral virtues which presuppose language and other characteristically human capacities and relations, such as honesty, veracity, promise-keeping, fairness, modesty as opposed to

> arrogance, and so on, as well as those detailed moral principles which vary from one society to another. (Mackie 1982a: 160–61)

Ethics, for Mackie, is thus the product both of biological and cultural evolution.

> We are, and will remain for the foreseeable future, what biological evolution over millions of years has made us. But on that practically unalterable basis, cultural evolution has already erected a far more complicated set of superstructures: we obviously do not behave *just* in ways that are biologically determined. (Mackie 1982a: 164)

But if pre-morality is the product of natural evolution, then why does Mackie say that morality solely consists of subjective values and human beings must invent right and wrong? If Mackie is correct that pre-morality *is* the product of natural evolution, then isn't there *some kind of* objectivity to morality? By Mackie's own reckoning, it seems that the process of natural evolution has put the main outlines of morality into place. If so, how much of it can we actually invent? There seems to be a tension in how Mackie brings error theory and evolution together. We see the same incongruity in Ruse's conjoining of error theory and evolution. According to Ruse, "If the empirical thesis . . . is right, there is no objective ethics. Nor can one readily see how the objectivist might patch up the situation, making his/her position compatible with evolutionism" (Ruse 1998a: 264).

Ruse has sought to bolster error theory with evolutionism. Yet, he says ethics is part of our human nature.

> Humans share a common moral understanding. This universality is guaranteed by the shared genetic background of every member of Homo sapiens. The differences between us are far outweighed by the similarities. . . . There is, therefore, absolutely nothing arbitrary about morality, considered from the human perspective. (Ruse 1998a: 255)

Ruse talks of morality as part of our human nature while at the same time claiming that morality's objectivity is an illusion. Would not the more straightforward interpretation be that, since morality is part of our human nature, then if perhaps human nature is objective, then morality is also objective? By closely linking ethics with human nature, it seems that Ruse himself has provided objectivists with all they need to "patch up the situation" and make objectivism compatible with evolutionism.

Further, in an effort to show that biology can indeed throw light on ethics, Ruse claims that "there is evidence from human studies pointing to uniformities of moral beliefs beneath all the cultural variations and that these uniformities are innate rather than learned" (Ruse 1993: 148). This seems to be the kind of information that would be relevant to assessing the argument from relativity. Whereas Mackie used the argument from relativity to support error theory, Ruse's observation about uniformities of moral beliefs beneath cultural variations severely weakens the argument from relativity. Mackie's argument is, if there really are objective values then why do we observe so much relativity? Ruse, by referring to data showing moral uniformity not relativity, opens up the possibility that perhaps objective values *are* motivating people to come together. This weakens the error theorist's claim that there are no objective values. If, under the veneer of cultural variations, ethics is not so arbitrary then why does Ruse say that ethics is a "collective illusion"? And why say, in Mackie's words, that we must "invent right and wrong"? There appears to be a tension between Mackie's notions that we create morality and yet that it has already largely been put in place by millions of years of evolution.[12]

One of the key issues in assessing error theory has to do with how proponents of error theory characterize commonsense morality. Not all philosophers believe that Mackie has accurately characterized it.[13] But even from Darwin's evolutionary perspective, there seem to be enough other elements of commonsense morality which make it reasonable to say that morality is ultimately grounded in human social instincts. Darwin, though he acknowledged an instinctive push toward ethics, is not led to then see ethics as error theorists do. Darwinian ethics, as characterized by Ruse, emphasizes that ethics is a cooperative strategy that is deeply rooted in the contingent nature of the human species. A Darwinian ethics would seem to vindicate commonsense morality by saying it is rooted in our genes, innate dispositions, and capacities, and that there are innate constraints on our human behavior. These features of Darwinian ethics do not seem to sit comfortably with error theory.

We may disagree on the merits of error theory itself and disagree about how *well* evolutionary considerations support the error theorists in making their case. But in terms of my overall project, these particular conclusions are not my chief aim. Amidst this disagreement about error theory and evolution there is meaningful discussion about evolution and the ultimate nature of ethics. This first chapter discussion, then, should allay any initial doubts about evolution and ethics being discussed alongside each other. Moreover, if error theorists can argue that an evolutionary perspective supports their theory about the nature of ethics, then there is logical space for other metaethicists, defenders of expressivism, relativism, or realism, for instance, to claim against error theorists that an evolutionary perspective supports *their* metaethical theory even more than it supports error theory. Mackie, Ruse, and Joyce see evolutionary thought as providing support for their metaethical theories. Can evolutionary thought better support other metaethical theories?

In the next chapter, I look at what evolution implies about the metaethical theory of expressivism.

NOTES

1. Mackie says that his view about the nature of ethics may also be called subjectivism or skepticism (1977: 17–18). Subjectivism is an apt term since if moral values are not objective, they are broadly subjective. But the term *subjectivism* has been used by other theorists to denote a different theory than the one Mackie is defending. Skepticism, too, is somewhat apt since Mackie is skeptical of objective values. But the term *moral skepticism* has also been used differently by different theorists. "Neither name is altogether satisfactory," he says (Mackie 1977: 18).

2. Mackie himself does not use the term *internalism.* And although *internalism* is a contemporary term, Mackie sees evidence for this aspect of ethics in modern and ancient characterizations of ethics. "Plato's Forms," Mackie says, "give a dramatic picture of what objective values would have to be," for Plato thought "that just knowing [the Forms] or 'seeing' them will not merely tell men what to do but will ensure that they do it" and "being acquainted with the Forms of the Good and Justice and Beauty and the rest, [the philosopher-kings] will, by this knowledge alone, without any further motivation, be impelled to pursue and promote these ideals" (Mackie 1977: 23–24, 40).

3. *Non*cognitivists, by contrast, maintain that moral judgments do not take truth values, so moral statements and value statements are *not* the kinds of statements that can be true or false. Noncognitivists claim that moral judgments and statements appear to be cognitive but their true function is noncognitive. The range of possible noncognitive functions includes prescribing and expressing. I discuss noncognitive ethical approaches in chapter 3: Evolution and Expressivism.

4. Another error theorist, Joyce, whose views I discuss in (§2.2b) prefers to define cognitivism and noncognitivism in terms of assertions, rather than in terms of truth values. Thus, in Joyce's view a cognitivist holds that to make a moral judgment is typically to make an assertion ("'a is F' . . . is typically used assertorically"), while a noncognitivist denies this (Joyce 2001: 8). For Joyce: "An error theory, then, may be characterized as the position that holds that a discourse typically is used in an assertoric manner, but those assertions by and large fail to state truths" (Joyce 2001: 9).

5. In the paper (1982b) Mackie builds on the work of Edward Westermarck (1862–1939) regarding retributive emotions and their importance in explaining how morality seems objective but truly is not. Mackie believes he was able to "supplement and round out Westermarck's theory, fitting his various theses into a more tightly argued and causally coherent account" (Mackie 1982b: 219). Westermarck had also developed a Darwinian theory for explaining the incest taboo. According to the political scientists Larry Arnhart, Westermarck's "biological explanation of the 'incest taboo' has recently emerged, after many years of neglect, as one of the best examples of how Darwinian social theory can explain human social behavior" (Arnhart 2001: 19). About the incest taboo, and Westermarck, see the papers in

Wolf & Durham (2004). We will look at the incest taboo and some of Arnhart's views in chapter 7.

6. Although Mackie (in 1977) clearly relies on motive internalism, in (1982c: 115) he endorses *both* motive internalism and reasons internalism. Note that motive internalism and reasons internalism as discussed here are varieties of *internalism about morality*, not *internalism about practical rationality*, which is a different, but related topic discussed in Williams (1980).

7. Mackie, in an article (1982a), does explicitly analyze some points about ethics from the perspective of Dawkins' model of evolution, but in that article Mackie does not discuss error theory. Dawkins responds favorably to one of Mackie's suggestions (1989: 321).

8. But Ruse nonetheless does not seem to support genic selection; he seems to endorse classical Darwinian natural selection that sees individuals as the units of selection (1998a: 17–18).

9. Earlier versions of this argument were put forward by Lewis (1947) and Plantinga (1993).

10. See Mizzoni (2013) for further critical discussion of Haught's argument.

11. Joyce entertains the possibility of a morality that relaxes its categorical "clout" by conceding its dependence on desires, but he is not convinced that it would be a genuine morality. He concludes that "practical clout really is a core desideratum of any moral theory" (Joyce 2006: 209).

12. For a more critical discussion of Ruse, see Mizzoni (2005). For a critique of Mackie's characterization of objective values and his case for error theory, see Mizzoni (1995). Darwall critiques error theory by arguing that one cannot rationally believe what one knows to be false. "It seems to follow, therefore," Darwall says, "that one cannot rationally believe the error theory and continue to hold ethical convictions as well . . . one can no more continue to *believe* that anything has value or disvalue, or is right or wrong, and believe also that these beliefs are false, than one can coherently think that the bentness of the stick is only an illusion and continue to believe that the stick is really bent" (Darwall 1998: 64).

13. Brink (1984: 114–15), for instance, says that Mackie is simply wrong about commonsense morality and motive internalism. For a critical discussion of how Joyce characterizes commonsense morality, see Mizzoni (2006).

Chapter 3

Evolution and Expressivism

When the police arrived at the Heaton home in Warwick, Rhode Island, in September of 1989 they found three mutilated bodies. Joan Heaton, 39, had been stabbed by a knife eleven times, strangled, and bitten in the face. Her daughter, Jennifer, 10, had been stabbed sixty-two times. Melissa, 8, had been stabbed eight times and her skull crushed. Fifteen-year-old Craig Price, who lived in the neighborhood, was arrested shortly after the bodies were found. When asked about the murders, Price shrugged his shoulders and responded: "Morality is a private choice." According to Police Captain Kevin Collins, who witnessed Price's confessions to the murders, "He just loves to kill. There's no doubt . . . that he's going to kill again" (Boss 1998: 97). If right and wrong is essentially a matter of how people feel and ethics is fundamentally about our attitudes to actions, events, and people, then is our judgment that Price's actions were immoral a function of how his actions produce negative feelings in us? And is Price's own judgment that his actions were moral "for him" a function of how his actions produce positive feelings in him?

As a first brief description of expressivism in ethics, in (§1.5) I said that expressivists hold that moral statements and judgments are not the kinds of claims that can be true or false, since expressivists view moral statements and judgments as expressions of emotion. In this chapter I first lay out the main points of expressivism as they have been developed by Simon Blackburn and Allan Gibbard.[1] The main ideas of this ethical theory include: expressivism, projectivism, quasi-realism, ethical language, moral sensibility, a logic of attitudes, and norm-expressivism. By using these central expressivist ideas, we will see how expressivists would explain the moral judgments involved in the case of Craig Price. Having outlined the main ideas of expressivism, I then discuss how Blackburn and Gibbard view the relationship of expressivism and an evolutionary perspective on human origins. Lastly, I will examine what Darwin's, Dawkins', Gould's, and Haught's models of evolution imply about expressivism in ethics.

3.1 THE EXPRESSIVISM OF BLACKBURN AND GIBBARD

3.1a Blackburn's Expressivism

Blackburn calls his theory about the nature of morality *expressivism*. Expressivism is the view that ethics is essentially about the expression of attitudes. Under this view, an ethical claim is an expression of attitude. Blackburn has been defending this theory of ethics, this "theory based on attitudes" in which ethical statements "express the speaker's attitude," since 1971 (Blackburn 1971: 123, 128). By *attitudes*, Blackburn means emotional dispositions and states of mind, items such as sentiments, desires, and emotions. In thinking about how human beings come to hold ethical views, Blackburn suggests that we think of the process in terms of an input side and an output side. On the input side are "the natural features of the world to which the subject is responsive in forming an ethical commitment" (Blackburn 1990: 198; 1998: 4–5). As a metaphysical naturalist, Blackburn views natural features as the only features of reality one could possibly respond to. Blackburn acknowledges that some ethical theorists have attempted to describe ethics as intimately connected to theism, but he holds that "the appeal to a supernatural order as a foundation for ethics is philosophically famously weak" (Blackburn 1995: 20). Blackburn wishes to account for the nature of ethics by making the least amount of metaphysical assumptions. If we take the natural features of the world as our source data, then with regard to ethics, Blackburn views projectivism as the most plausible ethical foundation. Rather than a metaphysically extravagant and speculative ethic, expressivism keeps ethics down to "the projectivist earth" (Blackburn 1985: 153).

With the natural features of the world on the input side, "the attitude or pressure upon action or choice" is on the output side (Blackburn 1990: 198). *Projectivism*, then, pictures the nature of morality in terms of "sentiments and other reactions caused by natural features of things, and we 'gild or stain' the world by describing it as if it contained features answering to these sentiments" (Blackburn 1985: 152). The nature of ethics comes to this: the world impinges on us and we develop attitudes toward it. Values and ethics are simply a complicated "network of attitudes" (Blackburn 1996a: 99). As a projectivist view about ethics, expressivism characterizes ethics as human projections of attitudes; all ethical values thus have a subjective source in that they derive from human subjects. Mackie, as we've seen in chapter 2, who defends an error theory in ethics, also views all ethical values as subjective (§2.1). But error theory and expressivism importantly diverge, as we will see below.

Blackburn asserts that projectivism or subjectivism alone cannot adequately explain or account for basic ethical elements. Notions such as ethical truth and falsity, surface features of ethical statements, moral reasoning, and moral facts initially seem at odds with a projectivist view of ethics, for when people make ethical claims, engage in moral reasoning, and refer to ethical truths, they don't overtly seem to be projecting their attitudes or making subjective value claims. This is what Mackie pointed out about commonsense morality: commonsense morality seems to presuppose that morality is objective and there are objective values. And ethical language behaves logically as if it were no different from non-ethical language.

Now if one were to take ethical truth and falsity, the surface features of ethical statements, moral reasoning, and moral facts at face value—as representing a moral reality that exists independently of human attitudes or projections—that would commit one to moral realism.[2] Mackie described the disconnect between metaethical subjectivism and commonsense morality in terms of an error, but Blackburn seeks to explain away this apparent disconnect with an explanatory program he calls *quasi-realism*. According to Blackburn, when "the projectivist adopts quasi-realism, he ends up friendly to moral predicates and moral truth" (Blackburn 1990: 206). Quasi-realism sets out to explain how attitudes get formalized or concretized into indicative statements that then get used in valid forms of moral reasoning. Because ethical language on its surface does not appear to only express attitudes, Blackburn supplements a projective account of ethics with a quasi-realist theory.

If we grant that quasi-realism can explain how expressions of attitude can take on logical trappings, then what explains how we come to have attitudes in the first place, in what Blackburn calls the output side of ethics? Here Blackburn appeals to the notion of a moral sensibility:

> The real essence of ethics lies not so much in the networks of propositions that it has people asserting, but in the underlying contours of their sensibility: the contours of attitude and emotion that drive them in their actions and their relations to others. (Blackburn 1996a: 85)

This moral sensibility, says Blackburn, can be improved, deteriorated, or defective (Blackburn 1988b). As an example of how he views the complicated dynamic between one's moral sensibility and one's attitudes, consider what he says about bull-fighting. Opposing bull-fighting, he says, is "a proper, necessary expression of an attitude toward our own attitudes" (Blackburn 1985: 153). He also says, "If everyone comes to think of it as permissible to maltreat animals, this does nothing at all to make it permissible:

it just means that everybody has deteriorated" (Blackburn 1985: 160). Blackburn does not believe that expressivism ends up being an "anything goes" theory of ethics, although he realizes that people might think his theory somehow weakens the status of ethics. Blackburn's response to such worries is that "when people fear that projectivism carries with it a loss of status to morality, their fear ought to be groundless, and will appear only if a defective sensibility leads them to respect the wrong things" (Blackburn 1985: 156). Elsewhere he calls it "a defective ethical sensibility" (Blackburn 1988a: 181). Blackburn thinks his ethical theory based on attitudes does not weaken the status of ethics because his notion of an adjustable network of attitudes that is regulated by fallible, yet tolerably reliable, moral sensibilities provides ethics with ample strength and durability. With regard to a serial killer like Craig Price, an expressivist will say that Price's moral sensibility has deteriorated or is defective. When Price judges that his killings are moral, he displays his defective moral sensibility. An expressivist will further say that when average citizens learn of Price's killings and judge his actions to be grossly immoral, these citizens are relying on their feelings and their properly functioning sensibilities.

Over the years Blackburn has defended expressivism and responded to objections to the theory. He has made efforts to make expressivism reconcilable with common forms of ethical language and ethical reasoning and to deal with the issue of embedded attitudes or "indirect contexts," which are logical puzzles that arise when expressions of attitude are inserted into logical arguments (Blackburn 1988b).[3] He has worked to develop a "logic of attitudes" to deflect these kinds of objections (Shafer-Landau & Cuneo 2007: 457). I will not go into the details of Blackburn's semantic proposals of "how a logic might be developed" to accommodate expressivism (Blackburn 1988b: 193). I will bring my summary of Blackburn's expressivism to a close, so that we may move on to view expressivism from an evolutionary perspective.

The following quote from Blackburn summarizes his expressivism:

> We should not theorize about morality and ethics as if they are in the business of describing aspects of the world—the moral and ethical aspects of the world. I think we should see their function differently. To enter a moral or ethical claim is to perform an action with some function like this: it is, amongst other things, to set oneself for or against something, to invite others to share this orientation, to prescribe courses of action, to lay down boundaries and give warning against trespass, or to smile encouragingly on conformity. It is to take up an attitude or stance, and centrally it involves making an emotional response to contemplated events and states of affairs. (Blackburn 1996a: 83)

About ethical concepts and ethical language more specifically, Blackburn says that:

> moral and ethical vocabulary is there primarily to articulate this network of attitude and emotion. It enables us to share our stances with others, to shape our sensibilities in light of the opinions of others, and to hear and voice assent and dissent. This is the core of the expressivist or projectivist position in ethics. The *essence* of ethics lies in its practical function. Ethical language is not there to describe further facts—the ethical facts—or to give a peculiar description of more ordinary natural facts, but to voice the responses we are to make to affairs. (Blackburn 1996a: 83)

Expressivism is an exhaustive theory about the nature of ethics, for as Blackburn says, "There is precious little surprising left about morality; its metatheory seems to me pretty well exhaustively understood. The difficulty is enabling people to appreciate it" (Blackburn 1985: 163).

3.1b Gibbard's Expressivism

At the outset of this chapter I mentioned that the main ideas of expressivism include projectivism, quasi-realism, ethical language, moral sensibility, a logic of attitudes, and norm-expressivism. I have saved the last notion—norm-expressivism—for this section, because it is not from Blackburn, but from Allan Gibbard's version of expressivism. Gibbard describes himself as providing an "expressivistic analysis" and views his theory of norm-expressivism as filling out Blackburn's quasi-realist program (Gibbard 1990a: 8; 2002: 55).[4] Gibbard refines expressivism, for there seems to be a bit more going on in ethical judgments than simply expressing one's emotions or attitudes. A straightforward example of expressivist analysis would have us believe that when we make ethical judgments we are expressing our emotions about something, and that is the most essential component of a moral judgment. But consider an example offered by Stephen Darwall, who asks: "Suppose you are emotionally overloaded at the end of a long day when you read about [a] terrorist slaughter. Under those conditions, you react not with outrage but with a barely stifled yawn. Does this commit you to thinking that the slaughter was boring rather than outrageous?" (Darwall 1998: 73). Darwall explains that this kind of situation doesn't actually pose a challenge for Gibbard's norm-expressivism. Gibbard's version of expressivist analysis says that in making ethical judgments one is expressing acceptance of a system of norms (Gibbard 1990a: 154). In making an ethical judgment, then, the warranted feelings may not be attendant, yet one can still hold that one *should* have those feelings. In discussing the emotionally overloaded individual mentioned above, Darwall explains that with norm-expressivism it is easy to account for the distinction between "the reaction he actually has to something and the reaction he *should* have—that is, the reaction he thinks the thing he is contemplating *warrants*" (Darwall 1998: 73). As we have seen with Blackburn's version of expressivism, expressivism does view ethics in terms of sentiments and feelings, yet Gibbard wishes to emphasize that ethics is also principally about norms and coordination. Sentiments are an integral part of the story but in order for sentiments to lead to social coordination, there is an irreplaceable need for norms. As Gibbard sees it, "Meshed emotions coordinate, and norms mesh emotions" (Gibbard 1990a: 299). For Gibbard, the feelings that are centrally relevant to morality are guilt and anger; thus, moral inquiry seeks to develop norms for these feelings (Gibbard 1990a: 291–92). Moral inquiry involves "deciding what norms to accept as governing moral sentiments" (Gibbard 1990a: 311). Moral norms are simply norms for the rationality of guilt and resentment (Gibbard 1990a: 47). In short, Gibbard's norm-expressivism views ethical judgments in terms of moral norms. Under norm-expressivism, to think something moral is to accept norms that prescribe it. To make a particular ethical judgment that an act is morally reprehensible, for instance, Gibbard explains, "is to accept norms that prescribe, for such a situation, guilt on the part of the agent and resentment on the part of others" (Gibbard 1990a: 47). Again, it is easy to see how Gibbard's norm-expressivism will characterize judgments about Craig Price's actions. Let us turn now to consider how Blackburn and Gibbard view evolutionary considerations as lending plausibility to expressivism as a theory about the ultimate nature of ethics.

3.2 EXPRESSIVISM AND EVOLUTION

Blackburn wishes to point out that an evolutionary perspective in no way undermines expressivism, but rather, puts expressivism in a favorable light. Gibbard makes an even stronger case for the connection between evolution and his theory of expressivism. He argues that evolutionary considerations *support* his ethical theory.

3.2a Blackburn on Expressivism and Evolution

Blackburn devotes an article (Blackburn 1996b) to answering some objections to Darwinism generally, which were volleyed by an anti-Darwinian.[5] In that article Blackburn tries to clear up some common confusions about Darwinism. He clarifies, for instance, that

> nothing in Darwinian theory allows you to say that because some pattern of behavior would increase the amount of some type of genetic material in future generations, therefore it will exist. It does not as it were allow you to say that whatever is right, is. Nor does it allow you to say that because some trait exists, therefore it is an adaptation, so that whatever is, is right. (Blackburn 1996b: 609)

Blackburn has also devoted several book chapters to clearing up some common confusions about ethics and evolution (Blackburn 1998, chapter 5.4; 2001, chapter 4). After all is said and done, it is clear that Blackburn views human beings as the products of Darwinian natural selection and he believes that the basic elements of expressivism fit very well into an evolutionary framework. In fact, he says, the evolutionary origin of human attitudes and moral sensibilities—key aspects of expressivism—is more than speculative, but supported by theoretical and empirical studies (Blackburn 1988a: 168–69). The expressivist relies on the concept of "attitudes"; Blackburn also describes them as "stances." By *attitudes*, Blackburn means emotional dispositions and states of mind, items such as sentiments, desires, and emotions. To make an ethical claim is to "take up an attitude or stance, and centrally it involves making an emotional response to contemplated events and states of affairs" (Blackburn 1996a: 83). For Blackburn, these attitudes and stances have their source in our evolutionary past. Some attitudes and stances have met with evolutionary success and some have not.

> The evolutionary success that attends some stances and not others is a matter of the behavior to which they lead. . . . Animals with standing dispositions to cooperate (say) do better in terms of other needs like freedom from fleas or ability to survive failed hunting expeditions by begging meals from others. (Blackburn 1988a: 168–69)

In terms of projectivism, Blackburn's view is that human beings (as natural beings) experience natural features of the world and project their attitudes toward them. As a thoroughgoing metaphysical naturalist, Blackburn would say that human beings are among the natural features of the world. Human beings, like all other living beings, are the natural products of the natural process of evolution, or natural selection. So when Blackburn says that we respond to the natural features of the world and then project, he will have to include aspects of human nature as a subset of those natural features of the world to which human beings respond.

A moral sensibility is basically how we come to have attitudes in the first place. For Blackburn, our moral sensibility has its roots in evolution. It doesn't come to us fully formed though, because Blackburn holds that a person's moral sensibility can be improved, deteriorated, or defective (Blackburn 1988b). As social and cooperative creatures we "share our stances with others, to shape our sensibilities in light of the opinions of others, and to hear and voice assent and dissent" (Blackburn 1996a: 83). An important means we have for doing so is language. As evolved creatures we have "moral and ethical vocabulary . . . primarily to articulate this network of attitude and emotion" (Blackburn 1996a: 83). We need this ability to articulate our ethical stances, since it enables us to share our attitudes. And in the process, moral sensibilities—ours and those around us—get shaped.

Moral discourse, although it is at bottom only about attitudes—doesn't *appear* to be about attitudes. The expressivist's quasi-realism and logic of attitudes are connected to evolution because a naturalist perspective needs to explain "why moral discourse should have all the grammatical and logical trappings" that it does; and the answer is furnished by evolutionary considerations (Shafer-Landau & Cuneo 2007: 36). As evolved creatures we need some way to share our attitudes with those around us. Standard grammar and logic allow us to do this.

3.2b Gibbard on Expressivism and Evolution

By Nicholas Sturgeon's estimation (Sturgeon 1992: 111), "Gibbard's . . . is by far the clearest and most carefully worked out of the accounts that begin with an evolutionary story about our moral sentiments and conclude with an irrealist view of moral thought and discourse." Sturgeon means this in comparison with Blackburn's account (Sturgeon 1992: 113). As Sturgeon puts it, Gibbard believes that his norm-expressivism follows from an evolutionary story "not deductively, but rather with the kind of plausibility characteristic of scientific inference" (Sturgeon 1992: 103). Gibbard is careful when discussing evolutionary and human motivation: he believes that "any link between evolutionary considerations and what it makes sense to want will be subtle and indirect" (Gibbard 1990a: 29). He admits that an evolutionary approach will be speculative, since "we ill understand our own evolution;" yet "try fitting together evolutionary theory and an account of common experience, and we may develop better maps of both" (Gibbard 1990a: 256). More strongly put, though, he says that "our innate propensities are a product of Darwinian evolution" (Gibbard 1990b: 790).

Gibbard wishes to distinguish expressivism from other metaethical accounts, such as moral realism, as well as to show how an evolutionary account seems to work *better* with expressivism than with moral realism. Viewing our innate propensities as products of Darwinian evolution is "a central part of our best naturalistic account of the world and our place in it," he thinks (Gibbard 1990b: 790). So, if expressivism has a better fit with evolution than other metaethical theories, then expressivism will seem more plausible. Moral realists, for instance, will want to say that moral facts and ethical truths represent a moral reality of some kind, one that exists independently of human attitudes or projections. In contrast with such a view, expressivists like Blackburn and Gibbard view ethical judgments not as judgments of fact at all (Gibbard 1990a: 107). Gibbard suggests that his speculative evolutionary account helps the expressivist's case over the moral realist's. If expressivism is correct, Gibbard says, "Our normative capacities can be explained without supposing that there is a special kind of normative fact to which they typically respond" (Gibbard 1990a: 107).[6] Gibbard sees his norm-expressivism as fitting very well with an evolutionary framework. The acceptance of norms is a central dimension of norm-expressivism, and the human ability to accept norms is bequeathed to us by evolution. As Gibbard puts it:

> The capacity to accept norms I portray as a human biological adaptation; accepting norms figures in a peculiarly human system of motivation and control that depends on language. Norms make for human ways of living, and we can understand our normative life as part of the natural world. Normative talk is part of nature, but it does not describe nature. (Gibbard 1990a: 7)

"Why do people talk?" Gibbard asks (Gibbard 1990b: 787). Just as Blackburn emphasizes the important role for ethical language for sharing attitudes and shaping moral sensibilities, Gibbard also sees ethical language as playing a practical ethical role. A crucial biological function of language, Gibbard says, is "to adjust the terms of social cooperation" (Gibbard 1990b: 787).

Because social coordination is so important for human survival, Gibbard suggests that connecting norm-expressivism with an evolutionary perspective is rather straightforward.

> The biological function of the mechanisms underlying our normative capacities is to coordinate. Hence the psychic mechanisms that produce normative judgments are not systems of natural representation, they are coordinating systems. Their biological function is not to put something in the head in correspondence with their subject matter; it is to coordinate what is in one person's head with what is in another's. (Gibbard 1990a: 110)

As an expressivist, Gibbard emphasizes that ethics is centrally about emotions and feelings. He understands emotions such as anger and guilt as "sentiments an evolved, highly social being might have" (Gibbard 1990a: 256).

> A highly social species like ours might well evolve to have the set of broadly moral sentiments we seem to observe in ourselves. These sentiments fall into no neat system: different ones at times push different ways, and even one by one some of them lack clear content. Sentiments of fairness and respect in particular come with a definite form but issue no fixed directives; their seeming content emerges from an intricate social dynamic. (Gibbard 1990a: 273)

Gibbard believes that we "attach a moral sentiment to certain outcomes of bargaining situations, and that the words involved are 'just' and 'fair'" (Gibbard 1982: 38). "To regard something 'as fair,' then," he says, "simply is to attach this moral sentiment to it" (Gibbard 1982: 38). In descriptions like this, we can see the close similarities to Blackburn's notion of projectivism. Also, in like fashion with Blackburn, Gibbard believes that "Darwinian forces shaped the concerns and feelings we know, and some of these are broadly moral" (Gibbard 1990a: 327). As with Blackburn, further, Gibbard believes the human moral sensibility can change, improve, become more mature, sophisticated, and complex. Gibbard writes:

> Up to a point we can be pleased with ourselves as well as horrified, and we can try to do better. An important part of reflective life is sorting out what really matters and why . . . . Some of the central elements of our moral makeup are things it would make sense to want—to want in oneself and one's fellows. (Gibbard 1990a: 327)

In summary, not only is Gibbard a well-known proponent of expressivism, but he views his version of expressivism as having close connections with an evolutionary perspective. Norm-expressivism, he thinks, plausibly follows from an evolutionary story: humans have evolved to have a set of broadly moral sentiments; they have an evolved capacity to accept norms; and ethical language has the important biological function of helping to coordinate social cooperation.

Let us now go further and view expressivism through several models of evolution.

3.3 MODELS OF EVOLUTION AND EXPRESSIVISM

3.3a Expressivism and Darwin's Model of Evolution

Contemporary expressivism is part of an ethical tradition that traces back to David Hume (1739, 1751). British moral philosophers of the eighteenth century frequently referred to the moral sense. Darwin was familiar with this tradition and he, too, frequently refers to the moral sense (Darwin 1871: 71, 73, 87, 93). Sympathy is another ethical concept from British moral philosophy that Darwin discusses. He approvingly quotes both Adam Smith (Darwin 1871: 81, no. 17) and David Hume on sympathy (Darwin 871: 85, no. 19).[7] Darwin acknowledges and emphasizes that human beings have feelings not only for themselves but for other human beings, most obviously the "parental and filial affections" (Darwin 1874: 41). According to Darwin, social instincts prompt feelings of pleasure and sympathy and acts of service (Darwin 1871: 72).

For Darwin, then, natural feelings of sympathy for others are aspects of the social instincts, which appear in animals and humans. Darwin says that "sympathy . . . forms an essential part of the social instinct, and is indeed its foundation-stone" (Darwin 1874: 42). Consistent with natural selection, Darwin observes that the strength of one's feeling of sympathy is variable (Darwin 1874: 58). Considerations like these may help to explain the behavior of someone like Craig Price. He seems to be severely lacking in his share of what Darwin calls social instincts, which normally include sympathies and affections for others. Sympathy is such an essential aspect of the social instinct, thinks Darwin, that someone who had no instinct of this kind would be "an unnatural monster" (Darwin 1871: 90).[8] In terms of *expressivism* and feelings of sympathy, the important thing about sympathy being bound up with the social instincts is that, for Darwin, the social instincts are the necessary ingredients for morality:

> Any animal whatever, endowed with wellmarked social instincts, the parental and filial affections being here included, would inevitably acquire a moral sense or conscience, as soon as its intellectual powers had become as well, or nearly as well developed as in man. (Darwin 1874: 41)

Darwin says he has tried to show that social instincts, intellectual powers, and habits lead to the golden rule, which he regards as lying at the foundation of morality (Darwin 1871: 106). The intellectual powers surely have a place of importance in ethics, as expressivists would agree, but for Hume and the expressivists, intellectual powers alone do not generate ethics. The most necessary ingredient, as the expressivists and Darwin agree, is sentiments and sensibility. Darwin seems to concur with Hume and the expressivists, then, that ethics is ultimately based on feelings. So Darwin's account does appear to lend itself quite well with the general shape of expressivism.

Darwin believes in moral progress: he describes how the standard of morality has risen higher in human history (Darwin 1871: 103). In passages of Darwin's book *The Voyage of the Beagle* (1839), he reveals his opposition to slavery; so it is reasonable to assume that he views the abolition of slavery as an example of moral progress (Rachels 1990: 23). Under expressivism, what could it mean to say that the standard of morality has risen higher and higher? The view that ethics is only about expressing our attitudes may seem to sit uncomfortably with the view that ethical standards have risen higher and higher. Yet expressivism, as developed by Blackburn and Gibbard, will say that the key here will be moral sensibility. Expressivists will likely seek to explain moral progress in terms of the improvement, maturity, and increased complexity of human moral sensibilities.

Blackburn calls expressivism and projectivism austere and slender, metaphysically, (Blackburn 1990: 208); and in that regard it seems that his approach to ethics would be agreeable to Darwin, since Darwin, too, seeks to account for human behavior and human ethical behavior only by using resources from natural history. It is obvious that Darwin does not go into as much detail about ethical language that expressivists view as involving issues of quasi-realism, a logic of attitudes, and norm-expressivism; yet Darwin does observe that language will have an important place in how humans acquire and develop a moral sense. In Darwin's estimation, once language has been acquired and a community's desires are expressed then this naturally provides a guide for actions (Darwin 1871: 72).

3.3b Expressivism and Dawkins's Model of Evolution

Like Darwin, Dawkins does not go into as much detail about ethical language as the expressivists do, but Dawkins has no trouble working into his model of evolution the basic outlines of expressivism—that ethics is essentially about feelings and attitudes and human moral sensibilities.

Dawkins holds that human moral sensibilities have evolutionary roots, and that our basic psychological attributes and tendencies are put in place by natural selection. He thinks that the well-understood processes of kin selection and reciprocal altruism can explain how human beings have evolved these sensibilities (Dawkins 1989: 191). These tendencies are simply built into human beings, Dawkins thinks. "Sensitivity for unfairness is built into us (e.g., how I feel when I pay taxes but others do not)" (Dawkins 2001: 10). Dawkins is comfortable with the idea that our moral sense, just as much as our sexual desire, is "rooted deep in our Darwinian past, predating religion" (Dawkins 2006a: 222). Because these roots are so deep, he thinks, "We should expect that research on the human mind would reveal some moral universals, crossing geographical and cultural barriers, and also, crucially, religious barriers" (Dawkins 2006a: 222). And Dawkins does see such evidence. He finds it very interesting that people can come to similar conclusions when faced with moral dilemmas.

> From the present point of view, the interesting thing is that most people come to the same decisions when faced with these dilemmas, and their agreement over the decisions themselves is stronger than their ability to articulate their reasons. This is what we should expect if we have a moral sense which is built into our brains, like our sexual instinct or our fear of heights or . . . our capacity for language. (Dawkins 2006a: 223)[9]

Dawkins seems to allow a kind of intuitionism, the notion that a sense of right and wrong is rooted deeper than our conscious decision-making processes. So, as he says, after citing some experimental work, people can sense that something is right or wrong without being able to articulate the reason *why* they think it is right or wrong.

This makes an interesting contrast with Dawkins' fellow genic selectionist E.O. Wilson. Wilson cannot tolerate any hint of intuitionism in moral philosophy. Wilson does indeed acknowledge the basic idea about evolution and moral sensibilities, for he says that "human emotional responses and the more general ethical practices based on them have been programmed to a substantial degree by natural selection over thousands of generations" (Wilson 1978: 6). Wilson hypothesizes that the process of natural selection "repeated through thousands of generations inevitably gave birth to the moral sentiments" (Wilson 1998: 253). Even in terms of ethical theory, Wilson recognizes that a theory of moral sentiments can easily be linked with an evolutionary perspective. He says that "a few investigators are now embarked on just such a foundational inquiry . . . reviving the idea of moral sentiments developed in the eighteenth century by the British empiricists Francis Hutcheson, David Hume, and Adam Smith" (Wilson 1998: 251).

Yet, the idea of ethical philosophers who consult "emotional control centers in the brain" and try to "intuit the standards of good and evil" is quite objectionable to Wilson. "What . . . made the hypothalamus?" he asks (Wilson 1980: 3). Wilson thinks that since natural selection is responsible for the brain and the moral sense, then that shows that intuitionism to be problematic. Here, Wilson seems too eager to wrest ethics from the hands of philosophers that he doesn't see, as Dawkins does, that if the moral sense *is* rooted deep in human beings due to natural selection, then it is predictable that human beings *would* intuit a sense of right and wrong. Because of this, wouldn't some version of intuitionism have prima facie plausibility?[10]

Other observations of Dawkins contribute to this kind of intuitionism. Dawkins points to a consensus about right and wrong, for instance. The fact that such a consensus about right and wrong can exist among humans may indicate that there is a deeply shared moral sense that people "tap into" through intuition. Dawkins holds that there is a consensus about what is considered right and wrong, and the consensus is surprisingly widespread, and has no connection with religion, in his view (Dawkins 2006a: 262).

By acknowledging the presence of a widely shared moral sense, Dawkins has set the stage for the possibility of moral progress. As we have seen with Blackburn, expressivists armed only with a conception of ethics as a network of attitudes can still uphold a belief in moral progress. They do this by envisioning a moral sensibility that gets developed and refined over long periods of time. Here is how Dawkins describes moral progress:

> Something has shifted in the intervening decades. It has shifted in all of us, and the shift has no connection with religion. . . . The shift is in a recognizably consistent direction, which most of us would judge as improvement. . . . Where, then, have these concerted and steady changes in social consciousness

> come from? How is it synchronized across so many people? . . . Some of us lag behind the advancing wave of the changing moral *Zeitgeist* and some of us are slightly ahead. But most of us in the twenty-first century are bunched together and way ahead of our counterparts in the Middle Ages, or in the time of Abraham, or even as recently as the 1920s . . . over the longer timescale, the progressive trend is unmistakeable and it will continue. (Dawkins 2006a: 268–71)

As a contemporary evolutionist, Dawkins does not view this kind of moral progress as a straight line that is predetermined by the unfolding of evolution. There is no teleology packed into his conception of evolution. Although there are ups and downs along the way (he mentions twentieth-century atrocities involving Hitler and Stalin, for instance), Dawkins simply observes that there is a broad trend of moral progress.[11]

3.3c Expressivism and Gould's Model of Evolution

Gould offers a naturalistic model of evolution, so there would seem to be a good fit between his model of evolution and expressivism, which is also naturalistic. In Blackburn's theory of expressivism, naturalism is easily apparent when we look at his concept of projectivism. The natural features of the world impinge on us and in response we form attitudes that we project onto the world as if the world contained those sentiments. In the following passage from Gould, he seems to adopt a similar way of looking at the situation:

> Nature, who is as she is, and who existed in earthly form for 4.5 billion years before we arrived to impose our interpretations upon her, greets us with sublime indifference. (Gould 1999: 177)

In §2.3c I mentioned that after Gould notes that a morality of absolute and unconditional categorical imperatives seems implausible, he then describes other types of moral principles, ones that are premised upon "desire, negotiation, and reciprocity," which have appeared in many cultures in many different eras, and have shown their wisdom (Gould 1993: 50). With this, he seems to accept the expressivist's main premise about ethics, since desires are usually thought to be a function of one's feelings and attitudes. Gould, then, is acknowledging that wise moral principles need to be based in nothing more than human feelings—the central expressivist tenet. In denying that ethics consists in unconditional and absolute ethical imperatives, but rather in principles based in desires and feelings, Gould stands in the broadly Humean tradition, the same tradition in which expressivists are situated.

With regard to the origins of these desires and feelings, Gould's hierarchical theory of selection—which pictures selection proceeding on various levels—has no difficulty incorporating the insights of kin selection, just as

Dawkins does, to account for how affections for those who are closely related to us get put in place by natural selection. Gould says,

> Our altruistic tendencies need not represent a unique overlay imposed by the demands of civilization. These tendencies may have arisen by the same Darwinian route via kin selection. (Gould 1977: 266)

Gould's theory of punctuated equilibrium would seem to provide support for the expressivist's view of a stable human nature with an implanted moral sense, "innate propensities," as Gibbard calls them. In the fossil record paleontologists observe that the great majority of species appear with geological abruptness and then persist in stasis until their extinction (Gould 2002: 749). Punctuated equilibrium says that this is how the macroevolutionary process unfolds. The theory thus supports the view that *Homo sapiens* is in stasis; there is a stable human nature. Gould says that human nature has been stable for tens of thousands of years and its continued stability is the most reasonable prediction (Gould 2002: 914).

> The last 50,000 years or more of human phenotypic stability becomes a theoretical expectation under punctuated equilibrium, and not the anomaly so often envisaged (and attributed to the suppression of natural selection by cultural evolution) both by the lay public and by many professionals as well. (Gould 2002: 79)

To get a sense of how Gould argues for this kind of point, consider what he says about speciation. In order for a new species to branch off from *Homo sapiens* the most likely way that this could happen is if there are isolated populations. But Gould doesn't think the level of isolation needed is at all possible, given the kind of global species *Homo sapiens* has become. Gould writes:

> So if speciation usually requires isolated populations, how (barring science fiction scenarios . . .) can a global species like *Homo sapiens*, endowed with both maximal mobility and an apparently unbreakable propensity for interbreeding wherever its members travel, ever expect to generate substantial and directional biological change in its current state? (Gould 2002: 913)

The expressivists' assumption about a stable human nature with propensities toward particular feelings and attitudes seems to be reasonable when viewed from Gould's evolutionary perspective.

3.3d Expressivism and Haught's Model of Evolution

In §2.3d I outlined what Haught's model implies about naturalistic accounts of ethics generally. He believes that a naturalistic ethics that tries to conjure the whole of ethics out of purely naturalistic materials without providing an explanation concerning the source of the native imperative to be responsible, will inevitably come up short.

In terms of expressivism, Haught doesn't seem to be a projectivist about value. Unlike Gould's comment above about how nature greets us with "supreme indifference," Haught, on the other hand, says that "the universe is not fundamentally indifferent or hostile to the realization of value; rather, it has always had an adventurous inclination to expand the provinces of beauty" (Haught 2000: 133). Haught believes that if one adopts a naturalistic view of ethics, then one must be committed to the view that human beings are the ultimate and sole "authors of our values" (Haught 2006: 162, 163). Naturalism is compelled to this, he says, since naturalists deny that there is an eternal source of goodness that exists independently from human beings (Haught 2006: 162). Haught finds this corollary of naturalism to be an indication of naturalism's implausibility, and he would say that expressivism suffers from the same deficiency. Because naturalism makes human beings the author of all values, naturalism "therefore cannot lead the intelligent and responsible subject to any secure foundations for either intelligence or responsibility" (Haught 2006: 166).

Haught does not view truth, whether moral or nonmoral, as something that could be subjective. He holds that "there is something imperishable about truth . . . since *we* will be gone, *where* will it be true?" (Haught 2006: 201). The obvious option for a theist is to say that God is the ultimate source and guarantor of truth. Haught says that theology has always identified God's mind as the repository of truth, and without such a repository truth could perish (Haught 2006: 201).

Blackburn holds that notions such as ethical truth and falsity, along with surface features of ethical statements, moral reasoning, and moral facts, will initially seem at odds with a projectivist view of ethics, because when people make ethical claims, engage in moral reasoning, and refer to ethical truths, they don't overtly seem to be projecting their attitudes or making subjective value claims. This is why Blackburn develops a quasi-realism, and logic of attitudes to fill out his case for expressivism. Haught, though, will point out that expressivism still ultimately rests on a subjective foundation and so falls short when accounting for the ultimate nature of ethics, since in his view "There is something imperishable about truth" (Haught 2006: 201).

In this chapter we have seen the concept of moral progress in Blackburn, Gibbard, Darwin, and Dawkins. Haught is another believer in moral progress. He will agree with the expressivists that moral progress is a datum that needs to be explained. Haught views the big picture of cosmic evolution as the playing out of a narrative cosmological principle: cosmic evolution is a long story, one that involves a God of love. Yet, Haught will not be persuaded by an expressivist explanation for moral progress, as it rests on nothing more than sensibilities and attitudes, which Haught thinks are too subjective to give enough substance as a plausible explanation of moral progress. Moral progress is also consistent with Haught's aesthetic cosmological principle, the universe's ongoing creation of beauty: as there is more complexity, there is more nuanced beauty.

For Haught, aesthetic and moral progress work hand in hand. In the biblical tradition, for instance, Haught sees the working out of the aesthetic cosmological principle. He says that Jesus' parables include marginalized members of society (Haught 2000: 134). He sees the promotion of diversity and inclusiveness as an example of harmonizing contrasts. Building an inclusive community that reaches out to minorities and the disadvantaged has an aesthetic dimension (Haught 2000: 134–35). Just as Dawkins notes a broad trend of moral progress over time, Haught will likely also point to a *broad* trend with regard to the biblical tradition, since episodes of violence and discrimination also occurred in that tradition.

Expressivists will look at these examples and say that the key here is moral sensibility. Expressivists will likely seek to explain the incorporation of the disadvantaged and the creation of a more inclusive community as simply the product of an improved, more mature, and more complex human moral sensibility that goes no deeper than refined feelings and attitudes. Haught will see the explanation of moral progress with reference to an improved network of attitudes as dubious.

3.4 CONCLUSION

Can expressivism fit within an evolutionary framework? If human biological evolution is part of our worldview, does an expressivist account of ethics make sense? Two recent defenders of expressivism, Blackburn and Gibbard, clearly think so.

Blackburn does not connect his theory of expressivism to evolution by going into great detail about how the main aspects of the theory (projectivism, moral sensibility, ethical language, quasi-realism, and a logic of attitudes) fit with an evolutionary perspective. Yet, there is enough there in his works—especially how human "stances" and "attitudes" (moral sensibilities) have evolutionary roots—to see that he believes that an evolutionary perspective does not in any way undermine expressivism but, rather, puts it into a more favorable light compared with other metaethical theories, such as moral error theory, moral relativism, and moral realism. Gibbard goes into much more detail about how he envisions expressivism as being tied with an evolutionary perspective. He speculates that "a chief biological function of our normative makeup is coordination" (Gibbard 1990a: 324). His norm-expressivism seeks to specify how ethics (as attitudes/feelings/sentiments) can contribute to the coordination of human social actions. He believes that norms are the indispensable lynchpin for how human beings achieve this.

Darwin does not offer a full-scale expressive theory about ethics. What he says about ethics leaves open several metaethical possibilities. "The imperious word ought," Darwin says, "seems merely to imply the consciousness of the existence of a rule of conduct, however it may have originated" (Darwin 1871: 92). Darwin isn't necessarily an expressivist, yet most of what he says about ethics does seem to fit into an expressivist framework: his emphasis, for example, when talking about ethics, on a moral sense, feelings, affections, and his view that human social instincts are the necessary ingredient for morality. Dawkins' model of naturalistic genic selection characterizes moral sensibilities as having been put into place by natural selection. Dawkins finds that a common moral sensibility is widely shared among human beings. He even observes moral progress. Gould's model of naturalistic hierarchical selection also works well with expressivism. Gould seems to adhere to projectivism about value; an ethical theory that says ethics is ultimately about the feelings and desires of human beings seems to be rather straightforward for him. Also, with his model of evolution there is enough of a stasis with regard to human nature that the status of ethics is not jeopardized by its only consisting of an elaborate network of attitudes.

As for Haught's model of theistic evolution and its implications for expressivism, he is clearly anti-naturalist and anti-expressivist. He views projectivism as a mistake. In contrast with projectivism, he views value actually *in* the universe, for example, in his employment of an aesthetic cosmological principle. He does believe there has been much moral progress, yet he would find the expressivists' attempt to account for it in terms of an incremental refinement of moral sensibilities to be unpersuasive.

Expressivists allow that our moral sensibilities can deteriorate, improve, or be defective. But Haught would likely be among those who think that expressivism ends up being an "anything goes" theory of ethics, a form of moral relativism. Not expressivists, though. Blackburn, for instance, says that "if everyone comes to think of it as permissible to maltreat animals, this does nothing at all to make it permissible: it just means that everybody has deteriorated" (Blackburn 1985: 160). Blackburn realizes that people might think that expressivism somehow weakens the status of ethics by characterizing ethics as only about emotions. But Blackburn thinks his ethical theory based on attitudes does not weaken the status of ethics, because his notion of an adjustable network of attitudes that is regulated by fallible, yet tolerably reliable, moral sensibilities provides ethics with ample strength and durability (Blackburn 1985: 156). Although expressivists like Blackburn do not maintain moral relativism, some ethical theorists do. In the next chapter I look at what defenders of moral relativism have said about the ultimate nature of ethics and how moral relativism is viewed from an evolutionary perspective.

NOTES

1. There are many variants of the metaethical position under review in this chapter. Of the many versions of this metaethical position such as emotivism, non-cognitivism, prescriptivism, quasi-realism, and norm-expressivism, defended by A.J. Ayer (1936), Charles Stevenson (1944), R.M. Hare (1952), Simon Blackburn (1993), and Allan Gibbard (1990a) respectively, what they all share in common is the claim that although moral discourse *appears* to be cognitive and behaves cognitively, at bottom moral claims are expressions of emotions, prescriptions, or attitudes. Each version of this metaethical position is in keeping with David Hume's (1739, 1751) notion that ethics is ultimately grounded in sentiment.
2. Moral realism is discussed in chapter 5.
3. This is known as the Frege-Geach problem. For more detail about how it poses a challenge to expressivism, see Sinnott-Armstrong (2000).
4. Blackburn, who must provide a logic of attitudes, believes that Gibbard, with his norm-expressivism, has an easier time generating "a smooth logic" (Blackburn 1988b: 195).
5. David Stove (1994).
6. Sturgeon (1992: 103; 1995: 417) doesn't believe that evolutionary considerations confer an advantage for expressivism over moral realism. I will pick up this thread of argument in chapter 5 that focuses on moral realism.
7. Darwin cites Hume's *An Enquiry Concerning the Principles of Morals* (1751), and Smith's *A Theory of Moral Sentiments* (1759). Making a point that applies to both Hume and Smith, Arnhart writes that "while Smith spoke repeatedly of nature as instilling those moral sentiments that would promote the survival and propagation of human beings as social animals, he could not explain exactly how it was that nature could shape the human animal in this way. Such an explanation was later provided by Charles Darwin" (Arnhart 2001: 15).
8. Yet Price may still have the ability to empathize with his victim's pain. John Deigh (1995: 761) describes the "troublesome case of sadistic pleasure." It's troublesome, says Deigh, because the pleasure the sadist gets is increased by the sadist empathically imagining his victim's pain. It might be odd to say that a sadist has empathy for his victim. But the oddness is alleviated by the realization that the sadist isn't empathizing with the victim as a whole person, but only with the victim's pain and suffering.
9. This claim of Dawkins may at first seem naïve. But consider an analogue in ethical theory. James P. Sterba argues that if we look hard enough we can even see a consensus among moral theorists (2005). Sterba (2005: 5) sees much more agreement among prominent ethicists than we may first imagine: "The relatively formal

tenets shared by Aristotelians [virtue ethicists], Kantians [deontological ethicists], and Millians [utilitarian ethicists] respectively, when correctly understood, do little to divide these groups from one another, at least at the practical level, with regard to the most morally defensible formulations of their views."

10. The fact that we have moral intuitions doesn't mean those intuitions alone *justify* our moral judgments. To use intuitions to try to justify our moral judgments, an intuitionist needs much more argument. But an intuitionist approach is not refuted merely by claiming that our moral sense has evolutionary roots. Moral intuitions and evolution will be discussed in §6.2.

11. In the extensive book *The Better Angels of Our Nature: Why Violence has Declined* (2011) Steven Pinker has attempted to document how over the course of human history violence has actually declined.

Chapter 4

Evolution and Moral Relativism

In some cultures, if a woman engages in pre- or extra-marital sex, she brings great shame to her family. The only way to erase this shame is for the woman to be killed by her brother, father, or husband. In this particular culture, this form of killing—so-called honor killing—is understood as morally required (Levy 2002: 179). Observations such as these lead some to the theory of moral relativism: all morality must be relative, since actions morally required in one culture are seen as extraordinarily immoral in another culture. Let us try to situate moral relativism among the metaethical theories under discussion in this book.[1] As we saw in chapter 3, *expressivists* claim that moral statements and judgments are not the kinds of claims that can be true or false, since moral statements and judgments are expressions of emotion. In (§1.5) I said that *moral relativists* claim that ethical statements and judgments *can* be true or false, and whether a particular ethical statement is true or false is always relative. *Moral realists* likewise claim that ethical statements can be true or false, and to that they add that some ethical statements and judgments are objectively true. According to *moral relativism*, though, there is no objectivity to ethics. Because metaethical subjectivism and metaethical relativism both deny the objectivity of ethics, they are often referred to as forms of metaethical skepticism.

In general, moral relativists say it is only meaningful to talk of right and wrong from within a particular set of folkways, conventions, or frameworks. A moral relativist will say that if I view honor killing as morally wrong, it is only because of particular values that are prioritized in the social framework that I subscribe to. Individuals with a social framework in which values are prioritized differently may see nothing immoral or unethical with honor killing, since that social framework with its particular network of values has historically developed through different folkways. If others do not share our conventions, then they have no moral reasons to refrain from violating our conventions.

The main ideas of moral relativism include: conventions, moral reasons, moral diversity, and moral disagreement. The two primary specimen examples of moral relativism I will look at in this chapter have been developed in the work of Gilbert Harman and David Wong.[2] Harman is a philosopher whose case for moral relativism relies on the notion that one's moral reasons are always tied to the conventions to which one subscribes (Harman 1977: 113). Wong, on the other hand, calls his version of moral relativism "pluralistic relativism" (Wong 2006). Wong emphasizes the difficulties in settling moral disagreements, as well as the limited constraints on what can count as a morality.[3]

Having outlined the main ideas of moral relativism, I will then discuss how some writers have characterized the relationship between moral relativism and an evolutionary perspective on human origins. Further, I examine what Darwin and Dawkins' models of evolution imply about moral relativism; and lastly I will consider what Gould and Haught's models of evolution imply about moral relativism.

4.1 THE MORAL RELATIVISM OF HARMAN AND WONG

4.1a Harman's Relativism

Harman, like the error theorists and expressivists we've looked at in previous chapters, defends naturalism. In Harman's view, though, a naturalistic approach to ethics most plausibly leads to moral relativism (Harman 1984: 30). Naturalists need to give an account of values in the natural world, and Harman thinks moral relativism does it best (Harman 1984: 29). For Harman, the issue is: "not what is compatible with the evidence, but what best accounts for it" (Harman 1984: 48). And for a naturalistic approach, moral relativism is simply most plausible.

In formulating his version of moral relativism, Harman uses the concept of "convention" or "framework." In one of his writings Harman says moral relativism is the view that "different agents are subject to different basic moral requirements depending on the moral conventions in which they participate" (Harman 1984: 30). In another work he says moral relativism is the view that "moral right and wrong (good and bad, justice and injustice, virtue and vice, etc.) are always relative to a choice of framework" (Harman & Thomson 1996: 3). According to Harman's moral relativism, even though some people may talk of "absolute rightness," there is really only rightness in relation to a particular framework (Harman & Thomson 1996: 17). Moral relativism rejects the idea that there is a moral law that applies to everyone (Harman 1984: 34–35). Moral relativism also

> denies that there are universal basic moral demands and says different people are subject to different basic moral demands depending on the social customs, practices, conventions, values, and principles that they accept. (Harman 1984: 35)

Under moral relativism, any individual who claims that something is "morally good, right, or just," is always making that claim relative to a particular moral framework (Harman & Thomson 1996: 62).

In Harman's version of moral relativism, he thinks it still makes sense to say that moral claims can be true or false; it's just that moral claims are always true or false in relation to a particular convention or framework. A moral relativist, Harman says,

> supposes that the truth conditions of moral claims are relative to one or another moral framework. But that is not to deny that these claims have truth conditions; and a moral relativist will suppose that it is often possible to find out that a given moral claim is true in relation to a given framework. (Harman & Thomson 1996: 158–59)

Moral relativism is a theory about the nature of ethics, but Harman's version also takes a position on the origins of ethics. Harman holds that "morality arises when a group of people reach an implicit agreement or come to a tacit understanding about their relations with one another" (Harman 1975: 84). Moralities are social and defined by the conventions of groups, Harman thinks (Harman 1977: 113). It might seem puzzling to say that one's morality is a function of one's group's conventions, since each individual belongs to different groups that will likely have different—perhaps conflicting—conventions. Harman doesn't think this is puzzling, though. If you belong to more than one group, and the different groups have different conventions, and you are wondering which conventions determine your moral obligations, Harman's answer is: "They all do" (Harman 1977: 113).

One of the reasons that Harman is confident about the relativist view that morality is defined by the conventions of groups is the sheer explanatory power of the theory. For Harman, "Our having the moral beliefs we have can be explained entirely in terms of our upbringing and our psychology, without any appeal to an independent realm of values and obligations" (Harman 1984: 32). "Moral conventionalism," as Harman calls it, is only one part of his theory of moral relativism; the theory also has other dimensions (Harman & Thomson 1996: 27). Harman also understands moral relativism to involve a rejection of the idea that there are *universally applicable objective moral reasons* (Harman & Thomson 1996: 63). His position is that an agent's *moral reasons* depend on the agent's desires, goals, aims, intentions, and values (Harman & Thomson 1996: 62). To illustrate the point, Harman uses several colorful examples: intelligent beings from outer space, cannibals, and criminals employed by Murder, Incorporated (Harman 1975: 85). In each of these cases, he says that if these agents have good reason to injure us, then it doesn't make sense that by their standards they ought not to harm us. With regard to the intelligent beings from outer space, he says, "There might be no reasons at all for a being from outer space to avoid harm to us" (Harman 1975: 87). By Harman's estimation:

> That a certain course of action on their part might injure one of us means nothing to them. . . . In such a case it would be odd to say that nevertheless the beings ought to avoid injuring us or that it would be wrong for them to attack us. (Harman 1975: 85)

A criminal, according to how Harman understands morality, "may well have no reason at all not to harm his or her victims" (Harman 1984: 36). The mere "possession of rationality," Harman holds, is not sufficient to provide a source for relevant reasons, since certain desires, goals, or intentions are also necessary (Harman 1975: 87).[4] As evidence of this, Harman observes that

> there are people . . . who do not act in accordance with the alleged requirement not to harm or injure others, where this is not due to inattention or failure to consider or appreciate certain arguments, or ignorance of certain evidence, or any errors in reasoning, or any sort of irrationality or unreasonableness, or weakness of will. (Harman 1984: 36)

Again, for Harman there are no universally applicable objective moral reasons; rather, an agent's moral reasons depend on the agent's desires, goals, aims, intentions, and values (Harman & Thomson 1996: 62). Harman is quite confident about this *moral reasons* argument for relativism (Harman 1984: 39).

Harman's rejection of the idea of universally applicable objective moral reasons fits with his moral relativist claim that

> there is no single true morality. There are many different moral frameworks, none of which is more correct than the others. (Harman & Thomson 1996: 5)

For Harman, there are no universal moral reasons because one's moral reasons are only meaningful against a particular backdrop. In addition to supporting moral relativism with consideration of moral reasons, Harman thinks *actual observed moral diversity* supports moral relativism. He says that the claim that there is no single true morality "is made very plausible by actual moral diversity" and "is a reasonable inference from the most plausible explanation of moral diversity" (Harman & Thomson 1996: 6, 8). Moral diversity, and the "apparent intractability of moral disagreements," provide evidence for moral relativism (Harman & Thomson 1996: 187).

4.1b Wong's Relativism

Wong's version of moral relativism is a bit different from Harman's. Wong does agree with Harman that "there is no single true morality" (Wong 2006: xv). And, like Harman, Wong thinks that a naturalistic approach to morality supports moral relativism (Wong 2006: xiv). Yet, while Harman's form of moral relativism with its moral conventionalism seems to imply that anything can count as morality as long as there are conventions in place, Wong seeks to develop a theory of "relativism with limits" (Wong 2006: xiii). Wong calls his view "pluralistic relativism" because it "recognizes limits on what can count as a true morality" (Wong 2006: xv). For Wong, these limits derive from the function of morality, and from what the relevant empirical theories tell us about human nature (Wong 2006: 44). In terms of morality's functions, Wong observes that morality has several functions; broadly two: an *intrapersonal* function and an *interpersonal* function. Morality functions to coordinate one's intrapersonal instincts, feelings, and desires; but morality also functions to regulate and organize interpersonal relations (Wong 2006: 43). In terms of human nature and psychology, Wong is confident that "the strength of self-interest in human nature necessitates a norm of reciprocity in moralities that effectively perform the interpersonal function" (Wong 2006: 47).

These kinds of "universal limits on adequate moralities" (Wong 2006: 65; 1991: 446–47) lead Wong to a rather moderate moral relativism. In the following passage we see Wong's considered judgment about the nature of morality. In passages like these, he does not sound like a moral relativist.

> Much of what is moral will be the same for, say, Asian and Western societies, because of the common functions of moralities, human nature, and similar conditions across human societies. The commonalities form a shared core that includes duties arising from special relationships, including duties to care for the young and to instruct them so that they can become full-fledged moral agents, norms of reciprocity, and the other norms and reasons necessary for accomplishing the functions of morality. (Wong 2006: 68)

Moral relativists usually emphasize moral difference, moral disagreement, and moral diversity; yet Wong's version of relativism acknowledges both "commonalities and differences in moralities across societies and within them" (Wong 2006: 68). Although he defends moral relativism, he has no qualms about recognizing "shared values" among those who are in moral disagreement (Wong 2006: 251), and "universal elements of adequate moralities" (2006: 58), and "a shared core" to morality (Wong 2006: 68). Wong believes that his theory of pluralistic relativism, with its notion of universal limits and constraints on morality, is not susceptible to the kinds of objections that apply to cruder versions of moral relativism. He writes:

> It is commonly thought that relativism simply regards the popularly accepted moral norms in a society as determinative of the truth conditions for moral statements in that society. This crude and uncritical conventionalism does not follow from pluralistic relativism, since moral norms are subject to evaluation according to the universal constraints on morality. (Wong 2006: 73)

Although Wong holds that there are "universal constraints on morality," which play a critical role for morality, nevertheless, he still thinks that there is enough diversity, and there are enough differences and disagreements—both among and within moralities—to say that morality is ultimately relative. For Wong, it is simply not the case that there is a single, true morality. Consider the contrast between "moralities oriented toward rights as a basic source of value" and "moralities oriented toward relationships and community" (Wong 2006: 22; 1991: 445–46). A rights-centered morality versus a community-centered morality is his paradigm example of moral diversity. Yet there are many, many other examples, he thinks. He feels that he paints a "portrait of moral diversity" that "points to the ubiquity of serious disagreement and therefore to the constant necessity for accommodation" (Wong 2006: 250).

In addition, Wong identifies a special kind of moral disagreement, a "widespread phenomenon" he calls "moral ambivalence" (Wong 2006: xiv). Moral ambivalence, he stipulates, is the "recognition of severe conflicts between important values and of the possibility that reasonable people could take different paths in the face of these conflicts" (Wong 2006: xiv). He later amplifies this definition and says that moral ambivalence is when "our sense of the unique rightness of our own judgments gets destabilized" when we come to "appreciate the other side's viewpoint" (Wong 2006: 5).

Moral ambivalence is not simply a kind of disagreement, but also a reaction to a disagreement. He says it is "a kind of moral disagreement" that "evokes a complex reaction I call 'moral ambivalence'" (Wong 2006: 5). As an example of moral ambivalence he points to the "stress and ambivalence we feel about conflicts between special duties associated with the personal perspective and duties owed to all arising from the impersonal perspective" (Wong 2006: 223).

As noted earlier, pluralistic relativism acknowledges commonalities and differences in moralities across societies and within them. As a kind of moral disagreement, moral ambivalence also "exists not only across different moral traditions but also within a single moral tradition" (Wong 2006: 23). Particular kinds of moral conflicts involve moral ambivalence because in these kinds of moral conflicts "we can understand both sides" (Wong 2006: 21). The reality of moral ambivalence, he thinks, "in conjunction with a naturalistic conception of morality, supports the conclusion that there is no single true morality" (Wong 2006: xiv).

Once we accept that moral ambivalence exists, Wong says that we must explain why it exists, and he thinks his theory of pluralistic relativism explains it best. Wong believes that "the strongest case for pluralistic relativism includes not only the claim that other moralities are as justified as our own but also the recognition of the genuine values that compose those moralities" (Wong 2006: 235). The following passage sums up the main elements of Wong's account of morality:

> My approach is explicitly committed to a naturalistic understanding of morality, and . . . the combination of moral ambivalence and a naturalistic approach generates a particular way of understanding why there is no single true morality but also of the basis for certain broad constraints on the range of true moralities. (Wong 2006: 98)

4.2 MORAL RELATIVISM AND EVOLUTION

In this section I look at how some writers have characterized the relationship between moral relativism and an evolutionary perspective on human origins, Harman and Wong included. William Graham Sumner (1840–1910), for instance, the pioneering sociologist, was known both as a defender of Darwinism and moral relativism. As a Darwinian, Sumner applied an evolutionary perspective to the study of society. He did much research in documenting the diversity he found in societies. He confidently viewed societies as organic entities with complexities that could only be understood by situating them in their historical context. For Sumner, social evolution was a direct development from biological evolution. The study of societies naturally led Sumner to the topic of morality. Having studied various moralities from various cultures, he reached the conclusion that morality is indeed relative. Morality is a product of "the folkways" (Sumner 1906: 41). For Sumner, it could not be otherwise: he was convinced that "mores define the limits which make anything right" (Sumner 1906: 438). In the extreme and indisputable variety of moralities he observed in cultures around the globe, he saw ample evidence for moral relativism.[5]

A number of contemporary thinkers, though, do *not* see moral relativism and evolution as a coherent fit. Michael Ruse and E.O. Wilson, for instance, when working with ethics and evolution, take a stance against moral relativism. Ruse and Wilson state clearly that an evolutionary perspective on human nature and ethics does *not* lead to moral relativism. Instead of ethical relativism, Ruse says that Darwinians recognize societal differences, but see them as modifications of a shared set of moral imperatives (Ruse 1998a: 255; see also Ruse & Wilson 1986: 188, 190).

Ruse continues:

> The differences between us are far outweighed by the similarities . . . I did not choose my moral code. For the Darwinian, the very essence of morality is that it is shared and not relative. It does not work as a biological adaptation, unless we all join in. (Ruse 1998a: 255)

As I said at the end of chapter 2, on error theory, Ruse, in an effort to show that biology can indeed throw light on ethics, claims that "there is evidence from human studies pointing to uniformities of moral beliefs beneath all the cultural variations and that these uniformities are innate rather than learned" (Ruse 1993: 148). There I said that this seems to be the kind of information that would be relevant in critically assessing Mackie's argument from relativity, which Mackie uses to support his error theory. Although Ruse and Mackie are fellow error theorists, Ruse's observation about the uniformities of moral beliefs beneath cultural variations weakens the argument from relativity. Recall that Mackie argued that if there really are objective values, why then do we observe so much relativity? Ruse, by referring to data showing moral uniformity, not moral relativity, opens up the possibility that perhaps objective values *are* motivating people to come together.

Whereas Sumner and most relativists emphasize moral diversity among and within societies, Ruse says a Darwinian must also admit the *similarities* between diverse moralities. Accordingly, says Ruse, "the evolutionary ethicist strongly denies relativity because then the universality of cooperating would break down" (Ruse 1993: 158). But, as we have seen (§2.4), Ruse's case against relativism also exposes tensions within his own version of error theory; because, if under the veneer of cultural variations, ethics is not so arbitrary, then why does Ruse say that ethics is a "collective illusion"?

The topic of moral relativism also came up in chapter 3 on expressivism. Blackburn the expressivist is adamant in denying that expressivism and relativism are one and the same, and denying that expressivism leads us to moral relativism. Blackburn seems to be in full agreement with Ruse and Wilson on this matter. "In the twentieth century," Blackburn writes,

> we have all been impressed by the diversities of human nature and human culture, but it is worth remembering some of the constancies that impressed earlier thinkers. We are social animals, with certain biological needs. We have to coordinate our efforts; we have to establish systems of property and promise-keeping and sometimes even government. (Blackburn 1996a: 90)[6]

Let's now look at our two specimen versions of moral relativism, Harman's and Wong's, and see whether their ethical theories cohere with an evolutionary perspective. Harman, in opposition to the naturalists just mentioned—Ruse, Wilson, and Blackburn—says that a naturalistic approach to morality lends itself to moral relativism. Harman's approach seems to fit with Sumner's general stance. Recall that for Harman's moral relativism, "Our having the moral beliefs we have can be explained entirely in terms of our upbringing and our psychology, without any appeal to an independent realm of values and obligations" (Harman 1984: 32). Such a view is reminiscent of Sumner's notion that morality is a product of folkways and that's the end of the story. Harman is aware of how biological considerations lead some theorists away from moral relativism. He acknowledges that "one possible view is that we are genetically constructed so as to feel such concern and respect for others and that makes morality possible" (Harman & Thomson 1996: 26). This sounds like what Ruse, Wilson, and Blackburn have said. Harman calls such a view the "sociobiological thesis" and says that there may be something to it (Harman & Thomson 1996: 26). In his considered view, though, he is not convinced; for he says, "There may also be a completely conventional explanation of our concern for others. Perhaps one develops concern and respect for others as part of accepting a convention, on the supposition that others are developing or have developed similar concern and respect" (Harman & Thomson 1996: 26). This, of course, is Harman's relativism, that relies on conventionalism. When comparing the "sociobiological thesis" to his own thesis, he oddly concedes that his hypothesis "is highly speculative, of course—just a possibility" (Harman & Thomson 1996: 27).

Harman's whole case for moral relativism doesn't merely rest on conventionalism, though. He also supports his case with his account of moral reasons, moral diversity, and moral disagreement. For Harman, these factors presumably tip the scales away from the sociobiological thesis.

Levy is another defender of moral relativism, and he takes head on the charge that biological considerations make moral relativism implausible (Levy 2002, chapter 5).[7] The kinds of biological considerations that Ruse takes as problematic for moral relativism include that morality won't work as a biological adaptation unless we all join in; that humans have shared basic values in addition to their differences; and the universality of cooperating. Levy doesn't think that these biological considerations "rule out the possibility of a significant relativism," because they still leave "room enough for important disagreements" (Levy 2002: 123). Moral disagreements, as we've seen, are central to the moral relativist's case. Also, Levy emphasizes that there is a gap between ethics and natural, biological constraints (Levy 2002: 137). Levy is making the common point that "we cannot simply read our values off our biology" (Levy 2002: 125). Levy's point is actually similar to Harman's position on moral reasons. Harman holds that there is no single true morality, and he defends this partly through arguing that there are no universal moral reasons. There are no universal moral reasons because one's moral reasons are only meaningful against a particular backdrop, the conventions an agent has agreed to. An agent's moral reasons thus depend on the agent's desires, goals, aims, intentions, and values (Harman & Thomson 1996: 62). The biological facts of human nature don't necessarily give an agent a moral reason to perform or refrain from an action; the agent is free to pursue myriad goals, regardless of his or her natural predispositions, even if those dispositions *are* biological adaptations. In short, Harman's account of moral reasons will always cut against any kind of argument that offers ethics as biologically based. If someone like Ruse counters that ethics is undoubtedly a product of human evolution and for that reason is therefore unlikely to be relativistic, Harman will remain unpersuaded. Since, for Harman, morality depends on the accepted conventions of agents who have subjective desires, Harman's version of moral relativism doesn't seem to acknowledge any constraints on what can count as a morality; so it doesn't matter to Harman that ethics is indeed the product of evolution. This is an aspect of Harman's version of moral relativism with which Wong disagrees (§4.1b). Here, it becomes clear that Levy's version of moral relativism is actually more similar to Wong's version, because both Levy and Wong accept that evolution has put particular constraints on morality. And Levy and Wong do not think these constraints stand in the way of moral relativism, because as Levy puts it,

> These constraints are very broad. Within the boundaries they set, there remains a great deal of room for an interesting descriptive relativism, for an endless variety of human moral systems, each of which fulfils the evolutionary function of morality well enough. (Levy 2002: 134)

Wong wishes for his version of moral relativism to be fully naturalistic, and he takes that to mean that our best account of morality should integrate well with "the most relevant empirical theories about human beings and society," and this includes an evolutionary account (Wong 2006: xiv, 36, 41, 42). Somewhat like Sumner, Wong says,

> Any system of belief that could be considered a morality today is the product of a long and complex evolutionary (biological and cultural) process. As a society's circumstances change, old practices and customs that regulate intra- and interpersonal conflicts die off or transform, and new ones emerge. (Wong 2006: 65)

Yet Wong argues that any adequate morality must contain "universal constraints on morality" (§4.1b), such as the functions of morality, human psychology, and the nature of human cooperation. Any adequate morality must acknowledge these constraints because they are a part of our evolved human nature.

4.3 MODELS OF EVOLUTION AND MORAL RELATIVISM

4.3a Darwin and Moral Relativism

Where do Darwin's claims about ethics stand with respect to moral relativism? Darwin does recognize that there are ethical differences from society to society, and also within societies, across time. Darwin observes that *some* ethical rules are grounded in "strange superstitions" (Darwin 1874: 59). When a rule held sacred by the tribe is breached, he says, it gives rise to "the deepest feelings" (Darwin 1874: 59). He marvels at "how so many strange superstitions have arisen throughout the world"; and he views them as arising "quite apart from the social instincts" (Darwin 1874: 59).

There are two other points Darwin makes that acknowledge ethical differences. First, when he says that any animal with social instincts and a well-developed intellect would inevitably acquire a moral sense, the moral sense that is acquired, he says, "wouldn't be exactly the same moral sense as ours" (Darwin 1871: 73). Second, he is confident that over time human sympathies have changed; they "became more tender and widely diffused, extending to men of all races, to the imbecile, maimed, and other useless members of society, and finally to the lower animals" (Darwin 1871: 103). Given his view of the close connection between ethics and sympathy, his observation about changing sympathies, then, amounts to an acknowledgment that differences in ethics emerge across time.

Harman takes diversity and intractability very seriously, as important evidence for moral relativism. While Darwin acknowledges moral diversity, such an observation doesn't lead him to relativism about ethics. Darwin seems to want to distinguish the ethical rules founded on superstition from those founded on social instinct. He would therefore seem to deny that ethics is solely the product of one's culture. Further, though, he says it is highly probable that "any animal whatever, endowed with wellmarked social instincts, would inevitably acquire a moral sense or conscience, as soon as its intellectual powers had become . . . developed" (Darwin 1871: 71–72). From this it follows that he doesn't see the inevitability of gaining a moral sense hitched to the contingencies of one's culture as simply a product of the folkways. Because Darwinian ethics denies that ethics is simply grounded in one's culture, since social and biological instincts are necessary, Darwinian metaethics is not relativistic. So, Darwin's approach doesn't align with Sumner's.[8] For Darwin, the moral sense has much to do with social instincts, so his view works well with "the sociobiological thesis," as Harman calls it, a view that Harman sees as antithetical to moral relativism, and one that Harman does not find convincing. There are reasons to think, then, that neither Harman's moral relativism nor Sumner's coheres with ethics as Darwin sees it through his model of evolution.

Harman's moral reasons argument for relativism may initially seem to take us back to Darwin's position against "formal morality," which I mentioned in §2.3a. A "formal morality" as Darwin describes it, is the view that moral actions are those "done deliberately after a victory over opposing desires" (Darwin 1871: 55). Joyce's error theory does characterize commonsense morality as presupposing that a moral agent is bound to the authority of objective moral reasons, regardless of his or her desires. Harman, though, does not characterize morality in this way, since Harman argues that an agent's moral reasons *depend on* the agent's desires. Harman's moral reasons argument therefore does not conflict with what Darwin says about formality.

Yet Wong's "relativism with limits," his "pluralistic relativism," which is a moderate moral relativism—a form of moral relativism that Levy also endorses—seems to be more fitting with Darwin's approach than Harman's. Darwin, too, seems to acknowledge what Wong calls "universal limits on adequate moralities." And what Ruse and Wilson say about evolution and relativist ethics seems to be in keeping with Darwin's model of evolution, even though, as discussed in chapter 2 on error theory, there are reasons to think that Darwin's model of evolution does not accord so well with error theory.

Both Harman and Wong, as moral relativists, say there is no single true morality for humans. Would Darwin agree? This is tricky. It is not the impression one gets from reading Darwin's thoughts on ethics. Darwin discusses various human characteristics that bear on "man's moral constitution." These include: a moral sense, a conscience, social instincts, parental and filial affections, virtues, sympathy, and other emotions like love, remorse, regret, and shame. Just like Ruse and Wilson do, Darwin seems to emphasize the similarity between humans, not diversity, when it comes to ethics. Darwin's approach seems to allow for a possible human moral objectivity. Maybe not the strong claim that there is one single true morality for all humans, but at least the claim that there are universal constraints, as Wong puts it. But rather than accepting these constraints and continuing to maintain moral relativism, as Wong does, it seems possible to acknowledge the constraints and point to them as substantially undermining moral relativism, as Ruse, Wilson, and Blackburn claim.

4.3b Dawkins and Moral Relativism

Viewing Dawkins' genic selection alongside moral relativism brings us back to §2.3b where I mentioned that Dawkins views genetic rules as constraining and directing behavior for the purposes of genes that are at the helm. Yet human behavior, by Dawkins' estimation, is not wholly directed by biological rules. Dawkins asserts that "among animals, man is uniquely dominated by culture" (Dawkins 1989: 3). Being dominated by culture, of course, carries cultural diversity with it. And, consistently, Dawkins acknowledges cultural diversity. He says that "human societies" admit of "astonishing variety" that suggests that "man's way of life is largely determined by culture rather than by genes" (Dawkins 1989: 164). Interestingly, it may be the case that other kinds of animals have their own cultural traditions. As the moral relativists do, Dawkins sees moral diversity *among* societies and also *within* societies. He says that "we have all changed massively in our attitude to what is right and what is wrong" (Dawkins 2006a: 265). To Dawkins, "There seems to be a steadily shifting standard of what is morally acceptable" (Dawkins 2006a: 268). Moral diversity and moral change are facts about which moral relativists demand explanation. Does Dawkins' acknowledgment of moral diversity lead him, then, to relativism about ethics? Is he led to something like Harman's conventionalism or Sumner's folkways?

It seems not. First, the moral change that Dawkins observes seems to him to be moral progress (§3.3b). He is confident that "over the longer timescale, the progressive trend is unmistakeable and it will continue" (Dawkins 2006a: 271). While moral disagreement and change are fitting with a relativist outlook on ethics, the notion of moral progress has been thought to be an obstacle to moral relativism (Stace 1937: 48–50). The challenge for moral relativists is to explain—or explain away—moral progress. If moral relativists wish to accept that moral progress has occurred, then they must explain with respect to what standard moral progress has occurred. If they wish to deny that moral progress is possible, then that flies in the face of the very observation that Dawkins makes about the progressive trend of social consciousness (Dawkins 2006a: 270–71). Further, Dawkins does not seem led to moral relativism, since he thinks there is evidence for moral universals, as well as a consensus about right and wrong, and a shared moral sense (Dawkins 2006a: 222–23). About morality, he thinks, "there is a consensus about what we do as a matter of fact consider right and wrong: a consensus that prevails surprisingly widely" (Dawkins 2006a: 262). And he is confident that "the consensus has no obvious connection with religion" (Dawkins 2006a: 262). More plausibly, in his view, our sense of right and wrong comes from our "Darwinian past," by which he means our evolutionary past, of course (Dawkins 2006a: 214). Dawkins claims that human moral sensibilities have evolutionary roots (§3.3b). "Sensitivity for unfairness," he says, "is built into us" (Dawkins 2001: 10). Like Darwin, he thinks that evolution is responsible for humans having a moral sense.

When it comes to morality, because Dawkins accepts that there is both cultural diversity and moral universals, his position does seem to be in keeping with Wong's version of moral relativism. Wong points to "universal constraints on morality," while at the same time claiming that there is enough moral diversity, difference, and disagreement—both among and within moralities—to say that morality is ultimately relative. Dawkins shares some of Wong's concerns about explaining morality, because he says that kin selection and reciprocal altruism do not do enough to explain cultural evolution and the immense cultural differences we see in the world (Dawkins 1989: 191).

Dawkins' suggestion for explaining the immense differences between human cultures is *memes*. Like genes, they are replicators. But memes are not biological, they are mental. Memes "propagate themselves in the meme pool by leaping from brain to brain via a process which, in the broad sense, can be called imitation" (Dawkins 1989: 192). Whether meme theory successfully explains cultural evolution is not our immediate concern here. With regard to moral relativism, the main point here is that as an evolutionist Dawkins, like Wong, views morality as possessing both universal and relativistic features. Wong has chosen to call such a view "pluralistic relativism." As with Darwin, though (§4.3a), the overall impression one gets from Dawkins' characterization of morality from an evolutionary perspective is not that of moral relativism. In the course of natural history, Darwin and Dawkins see moral progress, not unending ethical disagreement, and this seems to hold them back from an endorsement of moral relativism. I turn now to our two other models of evolution.

4.3c Gould and Moral Relativism

In Gould we find yet another example of a naturalistic evolutionist who does not think a naturalistic approach to morality lends itself to moral relativism. At one point, when discussing evolution and ethics, Gould even uses the phrase "destructive ethical relativism" (Gould 1999: 203–4). There, Gould is pointing out that his position does *not* lead to "a destructive ethical relativism," even though some might imagine that it *does*.

Before considering Gould's view on ethical relativism, let us recall what he says about human nature. As an evolutionist who defends a hierarchical theory of natural selection, Gould has some things to say about human nature, including the comparative influences of biology and culture on human beings. Gould writes:

> We have come to see ourselves as a learning animal; we have come to believe that the influences of class and culture far outweigh the weaker predispositions of our genetic constitution. Nonetheless, we have been deluged during the past decade by a resurgent biological determinism. (Gould 1977: 237)

For Gould, cultural forces swamp genetic constitution, and so with humans we cannot accurately posit a direct genetic basis for fundamental human traits, as he thinks biological determinists try to do (§2.3c). Interestingly, Gould agrees with Dawkins' claim, then, that humans are dominated with culture (§4.3b). Cultural variety is the norm, so Gould claims that human flexibility is what is most important for understanding and explaining human nature, not particular human traits that have been genetically determined (§2.3c). If so, why—when it comes to ethics—don't these considerations lead Gould to a moral relativism? Wouldn't the belief that culture dominates what humans are like, along with a belief in ineradicable human flexibility, help support the case for moral relativism?

The stasis of human phenotypic stability would seem to make the difference for Gould. Even though human beings are flexible and adaptable, and even though it is difficult to match *particular* human traits with a particular genetic basis, that doesn't take away a salient fact about human nature, namely that it has been in stasis for at least 50,000 years or more, as Gould says. For Gould, the reason for the stasis of human nature is not that cultural evolution has suppressed natural selection, however. Rather, on Gould's theory of punctuated equilibrium, human phenotypic stability is predictable (§3.3c). In chapter 3, while discussing expressivism I said that the expressivists' assumption about a stable human nature with propensities toward particular feelings and attitudes seems to be reasonable when viewed from Gould's evolutionary perspective (§3.3c). The expressivists' assumption about a stable human nature with propensities toward particular feelings and attitudes is actually part of the defense that expressivists such as Blackburn offer to distinguish their view from moral relativism (§3.4). There is reason to think that Gould would take the same approach against relativism. Gould would point out, like Wong does, that there are limits to what can count as an adequate morality, and these limits are partly set by what the relevant empirical theories tell us about human nature (Wong 2006: 44).

Gould says we cannot posit a direct genetic basis for fundamental human traits, so this would include human *ethical* traits, also (§2.3c). In terms of understanding values, Gould says we cannot look to nature, since nature is indifferent to humanity (§3.3c). We are simply unable to glean moral truth from the facts of nature, he thinks. But though we have this inability to glean moral truth from the facts of nature, he doesn't think this leads to a "destructive" moral relativism. As noted in §2.3c, Gould proposes that moral principles, those premised upon "desire, negotiation, and reciprocity," have appeared in many cultures in many different eras, and have shown their wisdom (Gould 1993: 50). These frequently appearing moral principles Gould mentions seem like those universals that Wong and Ruse take note of. One of the reasons that Ruse rejects moral relativism, for instance, is that he doesn't think it fits with the idea of universality of cooperating (§4.2). A recognition of these kinds of moral principles that Gould has in mind is not consistent with Harman's relativism, which denies that there are universal basic moral demands and universally applicable moral reasons (§4.1a). Wong, as we have seen (§4.1b), does recognize universal limits on adequate moralities, and he says these universal limits play an important role in evaluating moral norms and in helping us avoid crude conventionalism.

Wong believes that his theory of ethics that acknowledges these universal limits on morality still merits the name "moral relativism." But because Gould apparently thinks relativism is "destructive," he would not likely accept Wong's choice of label, even though, like Wong, Gould in addition to noting universal limits on morality, acknowledges the power of culture to shape human behavior.

4.3d Haught and Moral Relativism

Haught is another evolutionist who stands against moral relativism. He recommends that we avoid the "swamp of unqualified relativism . . . wherein religious and ethical claims would have no basis" (Haught 2000: 74). His model of evolution is theistic, so his tack will be that naturalistic approaches plunge us in the swamp of moral relativism. That naturalistic approaches land us up with moral relativism, interestingly, is what both Harman and Wong would actually say! Harman and Wong don't see this as cause for alarm, of course, since they don't view moral relativism as a "swamp."

Haught thinks that a naturalistic approach to ethics leaves us with ethical relativism as our only option because he claims that naturalists must be committed to the view that we are the authors of our values (§3.3d). As moral relativists like to point out, our values conflict quite a bit; so if we are the authors of our values, then moral relativism might very well be the truth about ethics. Haught himself, however, does not think that human beings *are* the ultimate source of value (§3.3d). There is something imperishable about truth and value, he thinks. So he concludes that it is more reasonable to say that the ultimate repository of truth and value is God (§3.3d). For Haught, theists have a secure foundation for values, naturalists like Harman and Wong do not (§3.3d).

It is common for theologians to reject moral relativism. While there are many different cultures, there is one God, they say. More specifically about moral relativism, Haught will focus on the relativist's interpretation of moral diversity and disagreement. Haught doesn't understand moral diversity and disagreement as moral relativists do. According to Haught's model of evolution, we should *expect* imperfection in the world, since the world is in process. Under Haught's model I suspect that moral disagreements and irresolvable moral dilemmas are viewed as imperfections of an imperfect world that is in evolutionary process. With regard to what Wong calls moral ambivalence, a particular kind of moral disagreement or attitude toward moral disagreement, Haught accepts that there will be ambiguity in an unfinished universe, so for him the fact that we feel ambiguity about moral situations doesn't detract from the overall moral arc of the universe. Further, Haught thinks that "we should not expect the world that a generous God calls into being to be instantaneously ordered to perfection," but we should instead expect to find a "divine allurement," which is "an enticing and attracting divine humility" (Haught 2000: 53). Haught is not a rigid absolutist when it comes to ethics, so what he would say about Wong's and Harman's claims that there is no single true morality is not a straightforward issue. If asked if there is a single true morality, he'll likely say "yes and no." No, because of the ambiguity and disagreement present, but yes because of the broad constraints on the range of true moralities, of the kind Wong speaks of.

Related to the "universal constraints on morality" are universally applicable moral reasons. Harman's version of relativism, as we've seen, denies that there are universally applicable objective moral reasons (§4.1a). Haught, though, would likely accept that there are universally applicable moral reasons. As in (§2.3d), under Haught's model of evolution, human conduct can contribute to the intensification (or deterioration) of beauty in the cosmos. And since the evolutionary process is an unfinished and undirected story, human action can make a difference in how the story unfolds and ultimately turns out. These two principles taken together would seem to provide the materials for filling in the objective reasons why human beings should engage in moral conduct.[9]

Another issue that Haught would press against moral relativism is moral progress. He believes there has been moral progress: it is part of his aesthetic cosmological principle, the universe's ongoing creation of beauty—as there is more complexity, there is more nuanced beauty (§3.3d). Moral progress is a notion seen as troubling for moral relativism (§4.3b), so Haught will also find moral relativism dubious for this reason. Before I proceed to the conclusion of this chapter there is one more issue worthwhile to consider. Wong (2006: xi) airs the complaint that moral relativism usually gets caricatured and is not fairly considered. Foot (1978: 152) and Harman (1984: 27–28) also say that moral relativism is not given the attention it deserves. Labeling ethical relativism as "destructive" or "a swamp" is the kind of treatment Wong regards as uncharitable. While some theorists have dealt with moral relativism in short order with swipes of this kind, I have tried to avoid this kind of perfunctory treatment. I have dealt with versions of moral relativism that relativists have actually defended, not straw versions of it. And I have devoted a whole chapter to moral relativism, incorporating passages from two prominent defenders of moral relativism, Harman and Wong.

4.4 CONCLUSION

Moral relativism is the third metaethical theory looked at so far. In this chapter I first explained the main elements of moral relativism by focusing on the versions developed by Harman and Wong. I sought to see whether moral relativism fits within an evolutionary framework. Harman does not connect his version of moral relativism to evolution, yet he does argue that naturalism leads most plausibly to moral relativism. In his version of moral relativism Harman stresses conventionalism. For Harman, we are all under multiple conventions. There is no single true morality, since there are many different frameworks. Harman believes that actual observed moral diversity and disagreement support his case for moral relativism. The very nature of moral reasons, he thinks, also supports moral relativism. This is because there are no universally applicable moral reasons: all moral reasons are relative to an individual subject and his or her desires, values, or goals. While Harman himself doesn't develop connections between evolution and moral relativism, he doesn't view biological considerations as having much relevance for ethics. More significantly, he thinks, humans operate in the space of reasons.

Wong goes into more detail than Harman about how to situate moral relativism within an evolutionary perspective. Wong agrees with Harman that naturalism leads most plausibly to moral relativism, but disagrees with Harman's moral conventionalism. In Wong's pluralistic relativism, or "relativism within limits," morality does, after all, have a function: an intrapersonal and interpersonal function, which constitutes universal limits on adequate moralities. Yet, he thinks there is enough difference and diversity to call his view moral relativism. Wong thinks his commitment to a naturalistic approach requires him to take empirical theories about human nature seriously; this includes evolutionary theories about human nature, which he views as his bolstering his case for the intrapersonal and interpersonal functions of morality.

Beyond Harman and Wong, other theorists such as Sumner, Ruse, and Wilson have commented on the relationship between moral relativism and evolution. Sumner conjoined moral relativism and evolution, but Ruse and Wilson saw them as incompatible. Because of the weight given to disagreement and diversity by relativists, and the weight Darwin puts on social instinct *over* culture, there are reasons to think that neither Harman's moral relativism nor Sumner's coheres with ethics as Darwin sees it through his model of evolution. Yet, Darwin's approach fits very well with Wong's "relativism within limits," since Darwin's approach seems to allow that there are universal constraints.

Dawkins, because he accepts that there are both cultural diversity *and* moral universals, also seems to be in keeping with Wong's version of moral relativism. Wong points to "universal constraints on morality," while at the same time claiming that there is enough moral diversity, difference, and disagreement—both among and within moralities—to say that morality is ultimately relative. Yet Dawkins sees moral progress, a notoriously difficult datum for moral relativism to accommodate. Dawkins' model could perhaps align with Wong's moral relativism if moral progress could be explained as moral progress made with respect to "universal constraints on morality." Because if moral relativists wish to accept that moral progress has occurred, they must explain with respect to what standard it has occurred.

Gould mentions that moral relativism is a potentially "destructive" moral view. It might be surprising to find that Gould, as an evolutionist, emphasizes cultural variety as most salient about human behavior. Human nature does have stability, though, too, so we also see potential for alignment with Wong's version of moral relativism. Gould accepts that there are some universal demands, so Harman's moral relativism seems too strong for Gould.

Haught stands against naturalism and the "swamp of relativism." Theism is a more secure foundation for values, he says. Harman and Wong claim that with a naturalistic account of ethics, moral relativism carries the day. Haught is well aware of such an inference and he'll press it, but he will also remind us that moral relativism does not capture the ultimate truth about ethics. For Haught, moral disagreements and differences are explicable without resorting to relativism as the explanation. Moral progress is a significant datum in need of explanation. Yet, Wong's relativism could perhaps explain moral progress in terms of "universal constraints on morality." Actually, if we look back at Haught's accusation that relativism is a swamp, he says that "ethical claims would have no basis" in a "swamp of *unqualified* relativism." If we *qualify* moral relativism, as Wong does by making provisions for universal constraints on morality, then perhaps that would then make moral relativism acceptable.

The next chapter (chapter 5) will be one more chapter on metaethics. In it, I will focus on moral realism, known for its attempts to establish ethics as objective. Moral realism is a metaethical theory that stands opposed to error theory, expressivism, and relativism. If humans are products of evolution, then how could ethics be objective?

NOTES

1. Some philosophers, such as Harman (1975, 1984) label the theory under discussion in this chapter "moral relativism," while others, such as Lyons (1976), use the expression "ethical relativism." I treat the expressions as synonymous.

2. It is best to distinguish between moral relativism and cultural relativism. *Cultural* relativism is usually understood to simply be the descriptive observation

that, as a matter of fact, different cultures have different practices, standards, and values. In this chapter, this view is referred to simply as moral diversity or cultural diversity. *Moral* relativism, though, is the more controversial theory that all ethical standards are relative to the degree that there are no permanent, universal, objective values or standards. Social scientists from the first half of the twentieth century have contributed to the linking in our minds of cultural relativism and moral relativism. But the distinction between cultural relativism and moral relativism is an important one to bear in mind when thinking through these issues. See Mizzoni (2009) for more on this distinction. Normative ethical relativism is not under discussion in this chapter. In this book I only consider live theoretical options. I don't see that theorists today defend normative ethical relativism. The easy objections to normative ethical relativism are rehearsed by Rachels (1999), and given a clear articulation in Stace (1937).

3. Richard Rorty is well-known as a moral relativist, yet he is not a self-described moral relativist. In different works, he has repeatedly denied that he defends moral relativism (Miller 2002).

4. There is a similarity to Joyce's argument for error theory that reaches a similar, though, different, conclusion (§2.2b). Harman writes that "further reflection along naturalistic lines suggests that apparent absolute values are often illusory projections of one's personal values onto the world" (Harman 1984: 42). He sounds like the error theorists Mackie and Joyce here. Yet, Harman is not content with such a quick characterization. Rather than emphasize an error at the heart of morality, he does want to talk about ethical standards and ethical truths that do indeed apply to one's actions. It's just that these standards and truths, he says, are a function of the conventions under which one's actions happen to be constrained; thus one's morals are relative to one's conventions. In the end, Harman endorses a moral relativism, not a moral error theory.

5. Some have claimed that Sumner's work does not offer "a completely consistent relativism" (Hofstadter 1944: 65; Shone 2004). This mostly has to do with Sumner's social Darwinism, a normative ethical view about what human beings should do given that the "progress of civilization . . . depends upon the selection process; and that in turn depends upon the workings of unrestricted competition" (Hofstadter 1944: 57). For Sumner, the obvious practical conclusion is to avoid "legislative meddling with natural events" and promote "laissez faire" (Hofstadter 1944: 62, 65). A completely consistent normative ethical relativism could not dictate *the* one true ethical principle to practically follow, however, since that would be inconsistent with the very idea of *relativism*.

6. Peter Singer's view on evolution on ethics, which I will discuss in chapter 10 on utilitarian ethics, accords with Blackburn's view that biological considerations make moral relativism implausible. Singer says: "While the diversity of ethics is indisputable, there are common elements underlying this diversity. Moreover, some of these common elements are so closely parallel to the forms of altruism observable in other social animals that they render implausible attempts to deny that human ethics has its origin in evolved patterns of behavior among social animals" (Singer 1981: 29). Singer also mentions reciprocity as one of many human universals that detract from relativism (1981: 37; 1999: 34 ff, 68–69). R.J. Richards (1986: 275) is also opposed to conjoining moral relativism with evolutionary ethics.

7. Jesse Prinz (2012) is another defender of moral relativism who employs an evolutionary perspective.

8. Sumner's "social Darwinism," also, is not in keeping with Darwin's model of evolution, or of Darwin's account of ethics. See note 5, above.

9. Harman would say—in the same way that Joyce critiques objective moral reasons (§2.3d)—that a given individual may have no reason to refrain from contributing to the deterioration of beauty in the cosmos. The idea here is that even if the person were to accept the plausibility of the aesthetic cosmological principle, the principle does not provide *this particular individual* with a reason to refrain from diminishing the amount of beauty in the cosmos. Haught would likely respond that what makes a principle or a reason objective in the first place is precisely that it doesn't matter that a particular individual doesn't subjectively and consciously possess the reason. What matters is whether it is possible for the individual to gain access to the reason and the principle; if it is not possible for the agent to come to have the reason then by the principle of ought implies can, the agent cannot justly be asked to abide by the reason. But if the reason *is* accessible to the agent, which in this case Haught says it is since it is predicated on his model of evolution, then it makes sense to say the agent is bound by objective reasons to fulfill his or her responsibility and contribute to the intensification of beauty in the cosmos, not to its diminution. Harman will not find this line of argument convincing. But my concern in this section is with what Haught's model of evolution implies about moral relativism; so, for my purposes, I am more interested in what Haught's model of evolution implies about Harman's moral relativism, than in what Harman's moral relativism implies about Haught's theory of evolution. Overall, I think it is safe to say that Haught's model of evolution does not prima facie lead us to suspect that moral relativism captures the truth about ethics.

Chapter 5

Evolution and Moral Realism

Are there moral facts and moral principles that are true, regardless of whether human beings believe in them?

A common objection to such a suggestion is that most of what we believe about anything, not just *morality*, can be traced to contingent historical sources. For instance, our believing that "the earth is many billions of years old, roughly round, not at the centre of the universe, and the site of millions of generations of evolutionary activity," seems to depend "on our living when and where we do," since, "a very small percentage of people across the ages have held such views" (Shafer-Landau 2003: 259). True, we learn of these facts about the earth through contingent historical processes, but that in itself does not necessarily undermine them *as* facts. The same may be the case with moral facts. Though we learn of moral facts through the historical contingencies of our lives, some people think that moral facts and principles may nevertheless exist on their own, somehow. Those who believe this, believe in a kind of moral realism.

As we have seen in previous chapters, expressivists claim that moral statements and judgments are not the kinds of claims that can be true or false, since moral statements and judgments are expressions of emotion. Moral relativists claim that ethical statements and judgments *can* be true or false, and whether a particular ethical statement is true or false is always relative. Moral realists, as I briefly mentioned in §1.5, also claim that ethical statements can be true or false; but, in addition, moral realists contend that some ethical statements and judgments are objectively true: they are true whether anyone *believes* them to be true.

In this chapter, first I outline the main ideas of moral realism, which include: moral reality, independence, moral facts, moral properties, and moral cognitivism. I look at two different version of moral realism, a naturalistic version (ethical naturalism), and a nonnaturalistic version (ethical non-naturalism). I begin with the recent and comprehensive defense of ethical nonnaturalism offered by Russ Shafer-Landau (2003). Shafer-Landau believes in objective moral standards. There are truths in ethics, he thinks, but the "fundamental truths in ethics seem to express metaphysical, rather than physical (or biological or chemical, etc.), necessities" (Shafer-Landau 2003: 59–60). He also holds that "ethics . . . is not a science" (Shafer-Landau 2003: 220). Because there is more to ethics than can be captured in physicalist, naturalist concepts, he thinks that ethics should be considered *non*natural.

Shafer-Landau's ethical nonnaturalism parallels non-reductionist theories in the philosophy of mind, which say that the mental is not identical to the physical. Such theories say that mental properties are not identical to physical ones, since mental facts are not physical facts. But, mental properties are realized by instantiations of physical properties (Shafer-Landau 2003: 72–73). As regards ethics, Shafer-Landau's nonnaturalism holds that "a moral fact supervenes on a particular concatenation of descriptive facts just because these facts realize the moral property in question" (Shafer-Landau 2003: 77).

After outlining Shafer-Landau's nonnaturalist moral realism I then outline Peter Railton's naturalist moral realism, which he has developed and defended in many articles, many of which have been collected in (Railton 2003). As a moral realist, Railton also believes in objective moral standards, moral facts, and moral truths. Yet he thinks ethics can indeed be reduced to physicalist, naturalist concepts. He defends a version of moral realism known as reductive ethical naturalism.

Once we have a sense of moral realism, we can then look at various recent attempts to link moral realism with evolutionary considerations. Following that, we will consider several evolutionary models alongside moral realism. The last section of the chapter serves as the conclusion of Part I of the book, which has focused on evolution and metaethics. With chapter 6, the chapter to follow this one, we begin to examine evolution and normative ethics.

5.1 NATURALIST AND NONNATURALIST MORAL REALISM

5.1a Nonnaturalist Moral Realism

G.E. Moore (1873–1958) is a well-known moral realist. Moore defended a *nonnaturalist* version of moral realism. William Frankena, a student of Moore's, summarizes Moore's realism as a thesis that says that "an ethical judgment is a true or false ascription to something of an indefinable and non-natural (or non-empirical) property" (Frankena 1957: 37). Moore's analysis of moral terms, moral language, and moral meaning led him to believe that moral properties are not, and do not reduce to, natural properties.[1]

Moore asserted—via his open-question argument—that any way we try to define a moral term such as "good," the definition is always subject to the further question, "but is *that* good?" (Moore 1903: 15–17). Although many philosophers were convinced that Moore had made things difficult for anyone who would suggest that moral properties could be reduced to natural properties, critics say a disadvantage of Moore's view is that he characterizes moral properties as nonnatural properties that we allegedly come to know through moral intuition. Shafer-Landau, though, 100 years later, comes to Moore's defense and offers a comprehensive recent case for nonnaturalist moral realism.

As a defender of moral realism, Shafer-Landau argues that there is a moral reality. When individuals make moral judgments, moral realists like Shafer-Landau believe these are best understood as referring to a moral reality. Shafer-Landau says that moral reality is "stance-independent" (Shafer-Landau 2003: 15). He writes:

> The way I would prefer to characterize the realist position is by reference to its endorsement of the *stance-independence* of moral reality. Realists believe that there are moral truths that obtain independently of any preferred perspective, in the sense that *the moral standards that fix the moral facts are not made true by virtue of their ratification from within any given actual or hypothetical perspective*. That a person takes a particular attitude toward a putative moral standard is not what makes that standard correct. (Shafer-Landau 2003: 15)

Shafer-Landau disagrees, therefore, with those who try to characterize ethics as the outcome of a social agreement of some kind, as ethical relativists do, for instance (Shafer-Landau 2003: 16). Shafer-Landau views ethics as independent of agreements. Moral reality is what it is, regardless of what humans think of it. In Shafer-Landau's view, then, the reason that most people have a sense that ethics is objective is simply that moral reality is stance-independent; ethics *is* objective.

> Moral realism is the theory that moral judgements enjoy a special sort of objectivity: such judgements, when true, are so independently of what any human being, anywhere, in any circumstance whatever, thinks of them. (Shafer-Landau 2003: 2)

As a moral realist, Shafer-Landau believes in objective moral standards. Moral facts are fixed by these standards, he says. As examples of moral facts, he mentions "Torturing anyone for pleasure is wrong" and "Prudent and principled opposition to dictatorship is right" (Shafer-Landau 2003: 165). He disagrees with naturalistic versions of moral realism that say that moral facts *are* natural facts. Moral facts are of a different order, thinks Shafer-Landau.

To explain how moral facts connect with natural facts he employs the concept of *supervenience*. He says there is a supervenience relation between moral facts and natural facts: "A moral fact supervenes on a particular concatenation of descriptive facts just because these facts realize the moral property in question" (Shafer-Landau 2003: 77).

Some moral realists claim that moral facts do causal work. Shafer-Landau does not think so. This is why moral facts and moral explanation do not compete with scientific ones, he thinks. Moral facts do supply reasons for action, however.

Shafer-Landau holds that moral properties, though constituted by natural properties, are nonnatural. He uses the "constitution" concept. He says that moral facts are not identical with natural facts, but moral facts are *constituted* by natural facts. In Shafer-Landau's account of ethics, then, moral facts and properties are constituted by and supervene upon natural facts and properties. Because moral facts and properties do not reduce to, and cannot be captured in physicalist, naturalist concepts, he thinks that ethics should be considered *non*natural.

Shafer-Landau believes there are truths in ethics, and so he espouses moral cognitivism. He thinks that when we make moral claims, they are true or false by reference to moral facts (Shafer-Landau 2003: 17). In his view "Moral facts are the truth-makers of true moral assertions" (Shafer-Landau 2003: 15).

I have only offered a very brief sketch of the main elements of Shafer-Landau's version of moral realism. Beyond the main elements of moral reality, stance-independence, moral facts, moral properties, and moral cognitivism, his systematic defense of moral realism contains many, many more detailed facets, as well as much careful argumentation for his metaethical views. But let us now turn to Peter Railton's naturalist version of moral realism.

5.1b Naturalist Moral Realism

Like Shafer-Landau, Railton holds that there is a moral reality (Railton 1986a: 171). Railton thinks that the natural and social sciences have drawn a picture of the world and our situation in it, and the picture incorporates various mechanisms by which we come to interact with each other and the world. Among these various mechanisms, is a "feedback mechanism" that ties moral reality to empirical reality (Railton 1986a: 172).[2] Railton does not see his version of moral realism as an extravagant cosmic moral realism (Railton 1986a: 201). He asks us to

> think of morality in a rather starkly practical—and perhaps therefore incomplete—way: as a set of behavioral restraints and recognized permissions and obligations that function within a group to promote co-ordination in those

circumstances in which co-ordination would be mutually beneficial. Nothing here about representing an independently-ordered reality. (Railton 1991: 187)

Yet, as a moral realist, Railton does not characterize ethics as the outcome of a social agreement, or as reliant upon a "world-wide consensus" (Railton 1986a: 195, 197). Railton regards his view as a moral *realism* because he views moral standards as independent of whether human beings believe in them (Railton 1986a: 172; 1992c: 47, 49). With this independence, there is a strong sense in which morality is objective (Railton 1998b: 142).

Although expressivists view ethics as primarily about attitudes, the realist holds that not accounting for the *grounds* of the attitude is a serious omission for a metaethical theory (Railton 1993a: 279). As Blackburn, who defends expressivism, poetically describes the situation, expressivism is the view that "moral features of things" are the "children" of our sentiments, while realists hold that the "moral features of things are the parents of our sentiments" (Blackburn 1981: 165).[3]

Railton refers to moral features of things as moral properties (Railton 1986a: 165). Like Shafer-Landau, Railton holds that moral properties are constituted of natural properties and supervene on natural properties.[4] One of the important differences between their versions of moral realism is that while Shafer-Landau's version characterizes moral properties as constituted by and supervening upon natural facts and properties, he says moral properties are not reducible to natural properties. Railton, by contrast, does think that moral properties are reducible to natural properties; thus Railton holds a reductive naturalistic moral realism.[5]

Railton clarifies what it means to say that moral properties are reducible to natural properties. His version of moral realism "reduces moral properties to complex social-psychological phenomena, not to extra-human Nature" (Railton 1993b: 325). He does not mean that moral properties are reduced to natural properties in the sense that the moral properties then become redundant or dispensable. That type of outcome may occur when *some* things are reduced (such as polywater, a form of water thought to have been observed in the 1960s), but that is not the case with ethics, he says.[6] With ethics, Railton claims, the reduction to natural properties does not have to *eliminate* moral properties, but can rather *vindicate* them (Railton 1989: 162).[7]

> To achieve a vindicating reduction of good, [we] must identify it with a natural property (or complex of properties) that, to a significant extent, permits one to account for the correlations and truisms associated with 'good'—i.e., is at most tolerably revisionist—and that at the same time can plausibly serve as the basis of the normative function of this term. (Railton 1989: 162)

Evidence that a property has been vindicated and not eliminated, is when the reduction does not strip the concept of explanatory role. For instance, let's say we ask, why did the radiator crack during the freeze? Having reduced water to H_2O does not strip the concept water of its explanatory role. One can either say the radiator cracked because it was filled with H_2O, or one can equally say it cracked because it was filled with water. Reducing water to H_2O is a vindicating reduction, not an eliminative one, and, further, reducing water does not strip it of its explanatory role (Railton 1993b: 327).

Similarly, in ethics, Railton claims, a reduction of moral properties to natural properties need not eliminate the moral properties. Neither must the reduction strip moral properties of having an explanatory role. With the reduction, actually, moral properties will be shown to connect with and be rooted in the natural world, rather than be an error (as error theory would have it), be the "children" of human sentiments (as expressivism would have it), or be a function of social conventions (as relativism would have it).[8]

Railton makes the case that moral facts can be seen as connecting to natural facts about human well-being. Moral facts are complex; they consist of primary values that pertain to basic human needs, and so we can regard our behavior as shaped and constrained by them (Railton 1986a: 199).[9] Railton says there are facts regarding the well-being of ourselves and others, from which we cannot claim exemption (Railton 1989: 155).[10] The well-*being* of an individual he actually sees as a *non*moral good. A *non*moral good or a *non*moral value is "that component of the good a life may hold for the person living it which is logically independent of the moral qualities of that person" (Railton 1989: 155).[11]

Railton shares Shafer-Landau's commitment to cognitivism, both believe there are truths in ethics. Expressivists take it as their task to show how essentially expressive language "could display on its surface thoroughly descriptive, cognitive grammatical and logical behavior" (Railton 1989: 153). About the expressivists' attempts to do this, cognitivists like Railton and Shafer-Landau think that "no entirely satisfactory reconstruction of this kind has yet been given" (Railton 1993a: 280).

5.2 Moral Realism and Evolution

In this section I look at how some writers have characterized the relationship between moral realism and an evolutionary perspective on human origins, Shafer-Landau and Railton included.

Some theorists claim that moral realism is fully consistent with an evolutionary account of morality (Sturgeon 1992: 112; Harms 2000; James & Carruthers 2008: 244; James 2009; Collier & Stingl 2013). James, for instance, has argued that when considering an evolutionary account of morality, moral realism is not only a possibility, but it does a better job of spelling out what the conditions for a morality that developed from an evolutionary past would have to look like (James 2009). Harms clarifies that the most plausible version of moral realism will not be a "species independent" moral realism, but one that is firmly rooted in the human species (Harms 2000). Collier & Stingl say that "moral values are a real part of the biological world, whether or not animals are able to perceive them" (Collier & Stingl 2013: 218).

Sharon Street (2006), though, says that moral realists have a tough time accommodating an evolutionary perspective. Shafer-Landau does admit that moral realism seemingly carries an extra burden as compared with expressivism, for instance. An attractive feature of expressivism is "the ease with which it comports with a prevailing scientific, naturalistic view of the world," and expressivism offers a picture much "simpler than that proposed by cognitivists, who are forced to augment the scientific world view with an additional layer of moral properties" (Shafer-Landau 2003: 21). But Shafer-Landau also points out that one of the *advantages* of cognitivism is that it "straightforwardly preserves ordinary talk of moral truth" (Shafer-Landau 2003: 23).

In Street's view, though, for moral realists to accommodate an evolutionary perspective they would have to admit that selective pressures have influenced evaluative judgments, such as "The fact that something would promote one's survival is a reason in favor of it" (Street 2006: 115). Her challenge to moral realists is that moral realists must explain the relation between these evaluative truths and selective pressures.[12] As we have seen, moral realists claim that evaluative truths are *independent* evaluative truths; and for Street that puts moral realists in a difficult position. If realists claim that there is no relation between selective pressures and evaluative truths then they face the possibility that our evaluative truths may be far "off track" (Street 2006: 122). If realists try to argue that selective pressures have tracked the truth, then, under such a view, the ability to track these truths was evolutionarily advantageous. But Street is skeptical of what a realist ends up with. About nonnaturalist realists like Shafer-Landau, Street says:

> To say that these [evaluative] truths could kill you or maim you, like a predator or fire, would be one kind of answer, since it makes it clear how recognizing them could be advantageous. But such an answer is clearly not available in the case of the independent irreducibly normative truths posited by the non-naturalist realists. In the absence of further clarification . . . the non-naturalist's version of the tracking account is not only less parsimonious but also quite obscure. (Street 2006: 131)

What makes things worse for the moral realists, Street thinks, is that there are other good scientific explanations of our evaluative judgments that are more parsimonious and less obscure (Street 2006: 155). "The best causal accounts of our evaluative judgements," she says, "whether Darwinian or otherwise, make no reference to the realist's independent evaluative truths" (Street 2006: 155).

But Shafer-Landau is clear that he does not view moral properties as causally effective. Shafer-Landau's suggestion is that nonnatural properties be thought of as perhaps similar to logical truths. That "laws require lawmakers" he says, "is only as plausible as the claim that all of the laws of logic, mathematics, physics, and chemistry are themselves products of construction. They may be, but surely the burden here is on the constructivist to show them so. These kinds of law certainly seem to be best construed realistically" (Shafer-Landau 2003: 45). If a tracking account works best for *these* truths, then it would also work for moral truths, even if the moral tracking account may seem more cumbersome and obscure than expressivist accounts that deny that our moral judgments track moral truth at all. As Shafer-Landau says, "if mathematical or physical laws do not always require lawmakers, then perhaps moral laws do not require lawmakers" (Shafer-Landau 2003: 46).

For Street, creatures who make particular kinds of evaluative judgments tend to survive; it doesn't matter whether those evaluative judgments are "true" or not (Street 2006: 155–56). But can the same thing be said about mathematical and physical judgments, that creatures who make particular kinds of mathematical or physical judgments tend to survive, and it doesn't matter whether those mathematical and physical judgments are true or not?

At first, Railton's reductive moral realism would appear to have an easier time handling Street's challenge, since he characterizes moral properties as in-principle reducible to natural properties. But Street thinks that when viewed from an evolutionary perspective, naturalistic moral realism runs into just as much trouble.

> The naturalist can certainly try to develop answers to these questions, but at least on the face of things, the prospects appear dim. Take the widespread judgment that one should care for one's offspring, for example. Exactly what natural fact or facts does the evaluative fact that one should care for one's offspring reduce to, or irreducibly supervene upon, and why would perceiving the natural fact or facts in question have promoted our ancestors' reproductive success? It seems unattractive to get into such complexities when one can just say, as the adaptive link account does, that ancestors who judged that they should care for their offspring met with greater reproductive success simply because *they tended to care for their offspring*—and so left more of them. (Street 2006: 131–32)

Although Shafer-Landau does not explicitly entertain how his version of moral realism might sit with an evolutionary approach, Railton does explicitly discuss how moral realism accommodates an evolutionary perspective.

Railton acknowledges that there has been distrust about bringing "evolutionary thought to bear on morality" (Railton 2000: 56). And he is familiar with attempts to use the theory of natural selection to debunk morality, though he is not convinced of them (Railton 2000: 56; 2010).

Railton views human beings as "the result of a particular evolutionary history" (Railton 1986b: 10). And he thinks that evolutionary thinking would suggest that there are "extensive similarities across individuals" (Railton 1998a: 115). When discussing "well-being," Railton says that our well-being is one of the things that matters to us most (Railton 1986a: 179). He sees this as expected from an evolutionary perspective, for "evolution will have favored organisms so constituted that those behaviors requisite to their survival and flourishing are associated with positive internal states (such as pleasure) and those opposed to survival or flourishing with negative states (such as pain)" (Railton 1986a: 179). Relatedly, when Railton describes the "feedback mechanism," which according to him, ties moral to empirical reality, since his version of moral realism is reductive, what this means is that there is a feedback mechanism that ties moral facts with natural facts. The most important natural facts upon which moral facts supervene, are natural facts about one's well-being. Railton grants that the feedback mechanism will have a considerable flexibility and degree of unreliability. As we can easily observe, humans today are primarily motivated by their wants, not their instincts. Yet, humans can conform their wants to their essential interests and we experience at least "enough functional success to hold our own in an often inhospitable world" (Railton 1986a: 181).

This kind of feedback mechanism does not, however, yield "universal historical progress toward worldwide consensus on moral norms" (Railton 1986a: 195). Other mechanisms in play, as well as social and historical factors, account for this (Railton 1986a: 194). Also, Railton thinks his theory would predict that over the span of history the scope of the moral community would expand, since the feedback mechanism would allow people to identify essential interests. He also observes that although moral principles have been described to have "various origins and natures," it has become more and more typical "to make some sort of intrinsic connection between normative principles and effects on human interests," again, as his theory would predict (Railton 1986a: 197–98). Lastly, in areas where the feedback mechanism would work best—areas where there are similar interests that are potentially infringed upon by others—there are widely agreed-upon prohibitions against aggression, theft, and breaking promises (Railton 1986a: 198). But in other areas in which there are "large asymmetries in the capacity to infringe upon interests," the feedback mechanism is still in place, though hampered; thus, we observe things like social hierarchies, slavery, or caste inequalities (Railton 1986a: 198). How might Railton use these basic concepts to respond to Street's challenge to the naturalist moral realist?[13] Street wants to know what natural facts the evaluative judgment "one should care for one's offspring" reduces to or supervenes on. Railton says that our well-being matters to us most. Our offspring are part of us. The well-being of our offspring matters to us. When our well-being and objective interests—understood as natural facts—are denied or frustrated, we feel that we should respond, since our well-being (and theirs) matters to us. The feedback mechanism that Railton speaks of is the tracking account that Street thinks a realist must employ. When Street asks why perceiving these natural facts would have promoted our ancestors' reproductive success, Railton's answer would be straightforward: when parents perceive that their children have objective interests that contribute to their children's well-being, and the parents realize that their own well-being and their children's well-being are entwined, then the parents may formulate the evaluative judgment "I should care for my children." Caring for the children will obviously contribute to reproductive success.

5.3 MODELS OF EVOLUTION AND MORAL REALISM

5.3a Darwin and Moral Realism

Bearing in mind the key elements of moral realism—moral reality, independence, moral facts, moral properties, and moral cognitivism—I will now explore whether Darwin's observations about ethics leave conceptual space for moral realism.

First, Darwin does not think that ethical standards are wholly derivable from one's particular culture, but thinks that ethical standards are more stably rooted. This implies that ethics, as Darwin characterizes it, has a degree of "independence." So, as mentioned in chapter 4 on moral relativism, Darwin's approach does seem to allow for a possible moral objectivity (§4.3a).

Can Darwin's model of evolution allow for the degree of independence necessary for a moral realism, though? To answer that question we should go back to Darwin and expressivism. In §3.3a I said that Darwin seems to agree with the expressivists that the most necessary ingredient in morality is sentiment and feeling, and that ethics is ultimately based on feelings. Expressivists hold that the moral sense is grounded in sympathy, sentiment, and feeling. Without the ability to feel, we are unable to experience morality, they say. An expressivist view of ethics does not allow for the degree of independence required for a moral realism. So if Darwin's metaethic is expressivist—which there is good reason to believe it is—then it is not realist.

A moral realist will respond, though, that even if, perhaps, the feeling of sympathy is the key that opens us to moral considerations, that in itself doesn't establish that expressivism captures the full truth about ethics. For, to say that feelings are the key that opens us to moral considerations does not establish that moral considerations *are* sympathetic feelings. If I need a radio receiver, for instance, to pick up radio waves, that doesn't establish that the radio waves are radio receivers. We're still talking about two different things: the perceiver and the thing perceived. A moral realist (naturalist or nonnaturalist) who holds that moral reality has independence from attitudes, agreements, and beliefs will say, then, that expressivists who maintain that ethical claims are expressions of attitudes, are collapsing the perceiver and thing perceived into each other.[14] A moral realist may simply say that Darwin, too, collapses ethical attitudes with ethical realities. The overall point here, though, is that Darwin's evolutionary approach does not foreclose the possibility of moral realism, even if Darwin himself favored a more expressivist metaethic.

Another issue that bears on moral realism and Darwin's evolutionary approach is moral progress. As mentioned in §3.3a, Darwin believes there has been moral progress, and he suggests possible reasons how and why it has occurred (Darwin 1871: 103). We observe moral progress because over time human "sympathies became more tender and widely diffused" (Darwin 1871: 69). With the expressivists, Darwin seems to hold that moral progress is the result of progressively refined moralities. Moral realism, however, is actually thought to more easily accommodate the notion of moral progress. Moral realists will claim that it is with respect to moral reality that we have seen moral progress: moral truths have been discovered, and moral facts and properties recognized. Under Railton's account, for instance, he says his version of moral realism would predict that "universal historical progress" is not *inevitable*; nevertheless, though, over the span of history the scope of the moral community would expand, since a feedback mechanism regarding essential interests would progressively stimulate people (Railton 1986a: 195). Over time, people come to appreciate the force of moral facts and properties, since as conscious knowers they are simply appreciating facts about their well-being and essential interests. Moral properties thus come to be known through the very same processes as natural properties come to be known.

Given Shafer-Landau's nonnatural account of moral realism, he will explain moral progress in a slightly different way. As a moral realist, he will likely say that expressivists who explain moral progress in terms of progressively refined moral sensibilities do not have the whole story. And as a defender of nonnatural moral realism, Shafer-Landau will likely say that Railton's naturalistic moral realism that emphasizes people coming to appreciate the force of moral facts and properties—which are facts about their well-being and essential interests—also is missing something. What is missing is reference to a nonnatural moral reality. Moral progress, Shafer-Landau will say, is simply gaining over time more knowledge about stance-independent moral reality.

Although Darwin's approach to ethics is decidedly naturalistic, that doesn't mean that his model of evolution is incompatible with Shafer-Landau's nonnatural version of moral realism. Evolution by natural selection is itself a naturalistic process. As mentioned above, Shafer-Landau points out that "if mathematical or physical laws do not always require lawmakers, then perhaps moral laws do not require lawmakers" (Shafer-Landau 2003: 46). Over millions of years, naturalistic principles of evolution throw up organic creatures that must conform to physical laws that those organic creatures themselves did not create. Shafer-Landau's gambit is that moral laws and principles may enjoy the same kind of objective status. Moral facts supervene on natural facts, but that doesn't mean that moral facts *are* natural facts. As there are principles of mathematics and physics, there may also be principles of ethics, and some of these principles—as they are nonnatural—come to be known because they can be self-evidently perceived by a knowing mind. As Shafer-Landau puts it, "Our knowledge of many moral principles derives from their self-evidence" (Shafer-Landau 2003: 246).

5.3b Dawkins and Moral Realism

As I discussed in §4.3b, Dawkins observes among human beings an "astonishing variety," which one would think might lead him to an ethical relativism; yet, because he accepts a shared moral sense, moral universals, and that there has been moral progress, his view of ethics as perceived through his model of evolution does not easily comport with ethical relativism. Dawkins writes, for example:

> If our moral sense, like our sexual desire, is indeed rooted deep in our Darwinian past, predating religion, we should expect that research on the human mind would reveal some moral universals, crossing geographical and cultural barriers, and also, crucially, religious barriers. (Dawkins 2006a: 222)

So in this chapter the question is whether Dawkins' characterization of ethics from an evolutionary perspective might fit with a moral realist view of ethics.

Dawkins observes a consensus about right and wrong that prevails surprisingly widely, that he thinks has no obvious connection with religion (Dawkins 2006a: 262).[15] What metaethical theory would make the best sense of this? Moral realism can capture it, since moral realists accept that there is a moral reality that is stance-independent.

Given Dawkins' commitment to metaphysical naturalism, he would not likely take nonnaturalistic moral realism seriously. Other scientists who write about ethics, such as E.O. Wilson, also take metaphysical naturalism as their starting point. Wilson, for example, writes:

> If the empiricist world view is correct, *ought* is just shorthand for one kind of factual statement, a word that denotes what society first chose (or was coerced) to do, and then codified. . . . *Ought* is the product of a material process. The solution points the way to an objective grasp of the origin of ethics. (Wilson 1998: 251)

So let's first consider Dawkins' model of evolution alongside Railton's naturalistic moral realism. Dawkins stops short of ethical relativism because of his observations about a moral consensus and moral progress; so one interesting thing to notice is that Railton says his version of moral realism does *not* rely on a worldwide consensus. For Railton, this means that even if there is a moral reality, it doesn't follow that we should perceive a world-wide consensus. Even though there is a feedback mechanism about objective interests that could account for the consensus and moral progress we see, other mechanisms work against moral consensus. These considerations actually reveal that there is no difficulty integrating Dawkins' approach with Railton's moral realism. Railton's tentative claims about moral consensus seem only meant to qualify his ethical theory and ward off any worries that his theory is naïve or optimistic.

Even though we would not think that Dawkins' approach would harmonize with Shafer-Landau's nonnaturalistic moral realism, let's look at the elements of Shafer-Landau's theory. Shafer-Landau has a non-reductionist theory of ethics, which he says parallels non-reductionist theories about the mind. So if Dawkins would agree that the mental is not identical to the physical, then perhaps Dawkins would agree with Shafer-Landau that the ethical is not identical to the physical either. In both cases, non-reductionists say that mental properties and ethical properties are realized by instantiations of physical properties; and in both cases mental properties and ethical properties supervene on physical properties.

Shafer-Landau thinks that all of the laws of logic, mathematics, physics, and chemistry are best understood realistically, not as products of human construction (Shafer-Landau 2003: 45). Dawkins is confident that natural selection works on this planet, and bets that "life all around the universe" would evolve by the process of natural selection (Dawkins 1983: 42–423). Presumably, if we could consider natural selection as a biological law, then both Shafer-Landau and Dawkins would agree that it too, should be understood realistically. Since the truth of mathematical and physical standards are "not to be explained by their having been accepted or created by anyone" and "just are correct," that means they are stance-independent (Shafer-Landau 2003: 46). So, again, for Shafer-Landau, "If mathematical or physical laws do not always require lawmakers, then perhaps moral laws do not require lawmakers," hence moral realism (Shafer-Landau 2003: 46).

But even if it is the case that the truths of moral laws are stance-independent, what would it mean to say that ethics should be understood nonnaturalistically? Most importantly, it would mean that moral facts and properties require more than empirical methods to be perceived. To explain how we come to know moral principles, Shafer-Landau uses the concept of self-evidence. He writes:

> I hope to make it plausible that our knowledge of many moral principles derives from their self-evidence. . . . For it isn't very plausible to suppose that our all-things-considered judgements about moral cases are themselves self-evident. (Shafer-Landau 2003: 246)

What he means by this is that we come to know moral principles and truths differently than we come to know scientific principles and truths. And the difference is that our knowledge of at least some moral principles derives from those principles being self-evident (Shafer-Landau 2003: 61). Shafer-Landau's account of self-evidence is that propositions can be self-evident even if some people don't find them obvious of believe them. "Self-evidence is a normative notion," he claims, since "self-evident propositions are those that are sufficient to *justify* belief" (Shafer-Landau 2003: 258). So the fact that someone does not believe it will not undermine the justification of the self-evident proposition.

So it even seems possible to integrate Dawkins' approach with Shafer-Landau's version of moral realism provided that one is willing to accept non-reductionism in ethics, and that ethics can be understood just as realistically as other disciplines, as well as a theory of self-evidence through which it makes sense to say that some moral truths and principles are self-evident. Let's now turn to Gould's model of evolution and moral realism.

5.3c Gould and Moral Realism

In the previous chapters we have looked at Gould's model of evolution alongside error theory, expressivism, and ethical relativism. With regard to those metaethical approaches, he seems solidly against ethical relativism, and while he might agree with aspects of error theory, he tends mostly toward expressivism.

Unlike error theorists, Gould does not think there is an error or illusion pervading ethics. But because he would take error theory's concept of patterns of objectification seriously—Mackie's explanation of why human beings would believe in objective values while there are none—Gould might find questionable the moral realist's claims that ethical statements and judgments can be objectively true, and that there are moral facts and moral principles that are true, regardless of whether human beings believe in them. In §2.3c I said that Gould would be receptive to the "patterns of objectification" explanation about ethics because of his deep appreciation for the contingencies of history and the power of culture to shape human belief systems. A moral realist, in an attempt to make patterns of objectification less plausible and to disabuse Gould of his leanings toward expressivism, would perhaps engage Gould with the opening vignette of this chapter concerning the contingent historical processes through which we come to learn the age of the earth, its place in the universe, and that there have been millions of generations of evolutionary activity. Moral realists emphasize that we learn of these facts about the earth through contingent historical processes, and point out that that in itself doesn't necessarily undermine them *as* facts. And the same may be the case with moral facts. Though we learn of moral facts through the historical contingencies of our lives, moral facts and principles may nevertheless be stance-independent, and just as objective as laws of logic, mathematics, physics, etc.

Since Gould opposes ethical relativism, he doesn't think his model of evolution necessarily rules out the possibility of objective moral standards and objective interests. But because of his commitment to naturalism, and because he finds the projectivist account of values that expressivism offers convincing, he might view the notion of moral reality with suspicion. His theory of hierarchical selection and punctuated equilibrium predicts a stable human nature for very long periods of time, which works well with moral realism. But the notion of a stable human nature works well with expressivism too, because of its reliance on sentiments and desires as foundational for a non-relativist ethics.

Gould does observe that many cultures have landed upon some similar ethical principles, which he sees as based on desire, negotiation, and reciprocity, but nothing deeper. Gould would need to be persuaded of moral facts and moral properties, since he prefers the simplicity of the projectivist/expressivist view. As Blackburn says, "Someone with realist sentiments makes the world rich, and the interpreting mind lazy; someone with the opposite instincts makes the world poor, and the interpreting mind busy" (Blackburn 1981: 181). Naturalists such as Blackburn and Gould see no need to make the world "rich" with moral properties and moral facts, since human minds and sentiments seem all that is necessary to explain ethics. Along these lines, Gould would also be skeptical of Shafer-Landau's nonnaturalist moral realism that uses the concept of self-evidence to account for how we come to know objective moral principles. But as I said above with Dawkins, in §5.3b, it might be possible to integrate Gould's approach with Shafer-Landau's version of moral realism if one is willing to accept non-reductionism in ethics, and that ethics can be understood just as realistically as other disciplines, as well as the possibility of at least some self-evident moral truths and principles.

5.3d Haught and Moral Realism

In previous chapters we have looked at Haught's model of evolution alongside error theory, expressivism, and ethical relativism, and saw how he rejects ethical relativism outright and also rejects the main components of error theory and expressivism. The underlying reason why all three metaethical theories come up short for Haught has to do with his rejection of naturalism. But, specifically, against error theory, Haught's model of evolution does not lead us to suspect that commonsense morality is in error or is simply a useful fiction. About expressivism, Haught rejects a projectivist account of value, and doesn't think that sentiments and sensibilities are sufficient to explain the nature of ethics. About relativism, Haught thinks ethics has a foundation and a basis and ethical relativism cannot really provide one. And he thinks the facts of moral diversity and disagreement should not lead us to think ethics is relativist at bottom, and the observation of moral progress shouldn't lead us to think ethics is ultimately relative, either. Given what we have seen of Haught's characterization of ethics, it might be fairly obvious that he holds a version of moral realism. It should also be obvious that he wouldn't see a *naturalistic* moral realism—such as Railton's—as viable. As mentioned in §3.3d, in Haught's view, naturalists deny that there is an eternal source of goodness that exists independently from human beings (Haught 2006: 162).

And therefore, "naturalism is compelled to insist" that "we humans are the authors of our values" (Haught 2006: 162). Haught is convinced that a naturalistic account of ethics, of any variety, cannot accommodate objective values.

This is a claim, though, that Railton will not accept. In response to Haught's charge, Railton would simply say that there is no good reason to make this claim about a naturalistic moral realism, because as a naturalistic moral *realist*, Railton maintains that there are moral standards that are what they are independently of whether humans recognize them or not. Under Railton's account, basic human needs and objective interests are the grounds of objective (nonmoral) values. For Railton, "Moral facts . . . consist of primary values that pertain to basic human needs" (Railton 1986a: 199; 1993b: 318). And moral facts and properties differ from natural facts and properties, yet moral properties are ultimately reducible to natural properties, and the reduction can vindicate moral properties, not eliminate them. Railton, as a cognitivist about moral language, holds that moral judgments and statements can be true or false, and some of them are objectively true. As these components are at the very heart of Railton's metaethical theory, he will not take lightly Haught's claim that a naturalistic moral realism is doomed at the outset.

Turning to Shafer-Landau's version of moral realism, from the perspective of Haught's model of evolution, the main components of Shafer-Landau's moral realism tie in easily, including Shafer-Landau's characterizations of moral reality, stance-independence, moral facts, moral properties, and moral cognitivism.

Shafer-Landau regards his version of moral realism as nonnaturalistic, and by that he means that moral standards are not reducible to other kinds of standards (Shafer-Landau 2003: 47). Moral reality is logically distinct from natural reality; moral facts and properties supervene on natural facts and properties. Ethics is not science. This version of moral realism is welcome to Haught's theistic version of evolution simply because it leaves logical space for more than naturalistic physical properties. Of course Haught fills out his version of nonnaturalism differently, since Shafer-Landau, in his defense of nonnatural moral realism, says he wants "to remain neutral here on theological matters" (Shafer-Landau 2003: 76). But for both Shafer-Landau and Haught, to make a true ethical judgment refers to moral reality, and they will both agree that there is something imperishable about truth, yet Haught will fill this out with a particular theory of truth, one that views God as the source and guarantor of all truth.

Haught will agree with Shafer-Landau's non-reductive approach both to ethics and the mind. In §2.3d I described how Haught is an evolutionist about the human mind, in that he believes that our capacities to know and understand are the product of the earth and evolution. Yet he still believes that subjectivity and critical intelligence are not explicable on purely naturalistic grounds. Thus, we could say he is a non-reductionist about the mind. With regard to ethics, Haught develops a similar position: many of our ethical capacities and characteristics are the product of our natural evolutionary history, yet there is a fundamental and spontaneous act of cognition to "be responsible," which is not explicable and not reducible to purely naturalistic dimensions of the human person.

Haught will also agree with Shafer-Landau's description of moral facts not only as nonnatural, but as providing reasons for action. Against ethical relativists, we saw that Haught will argue that objective reasons for actions are possible (§4.3d). For Haught, objective moral facts provide objective moral reasons, even in an unfinished and undirected evolving universe.

5.4 CONCLUSION

Chapter 5 began with the question of whether there are moral facts and moral principles that are true, regardless of whether human beings believe in them.

The metaethical theory of moral realism affirms such a proposition. Moral realism centers on the concepts of moral reality, independence, moral facts, moral properties, and moral cognitivism. After sketching two different versions of moral realism, a naturalist version defended by Peter Railton, and a nonnaturalist version defended by Russ Shafer-Landau, the chapter reviewed what various authors, including Nicholas Sturgeon, Scott James, and Sharon Street, have said about the relationship between moral realism and evolution. Following that, the chapter looked at moral realism from the perspective of four models of evolution. We have found that both versions of moral realism have the resources to respond to sharp critics of moral realism such as Sharon Street, and that moral realism remains as a suitable metaethical candidate that comports with an evolutionary perspective on ethics.

Part 1 of this book has focused exclusively on metaethical theories—error theory, expressivism, moral relativism, and moral realism. Our examination of error theory and evolution has determined that there are tensions and incongruities in the error theorist's case, and that evolutionary considerations actually help to bring those tensions to the surface. With regard to expressivism, we have found that although it does not fare well on Haught's theistic model of evolution, it does comport well with naturalistic models of evolution.

About moral relativism and evolution, we have seen that Wong's version of moral relativism that eschews moral conventionalism has the best fit with an evolutionary perspective. And finally with moral realism and evolution, we have seen that both naturalistic and nonnaturalistic versions of moral realism are suitable metaethical candidates that comport well with an evolutionary perspective on ethics.

Metaethical theories are theories about the general nature and origin of ethics. In the next chapter, we will begin our examination of evolution and normative ethical theories. Metaethical theories, as merely general theories about the nature and origin of ethics, do not provide practical guidance, while normative ethics do. Normative ethical theories offer direction about what we should do and what kinds of persons we should strive to be. How will evolutionary considerations impact normative ethical theories? We will look at several well-developed traditions of normative ethics, and then examine them from various evolutionary perspectives.

NOTES

1. Henry Sidgwick (1838–1900), another British moral philosopher, was also a proponent of this view (1874: 32–33).

2. See also (Railton 1993b: 318). One triggering mechanism is observed when certain "objective interests" (a concept Railton uses) are denied (1986a: 175).

When a group of people experience their interests and needs being denied by another group of people, for instance, pressure for social change will build up until the interests get satisfied.

3. Or again, similarly, "Someone with realist sentiments makes the world rich, and the interpreting mind lazy; someone with the opposite instincts makes the world poor, and the interpreting mind busy" (Blackburn 1981: 181).

4. As an example to explain the concept of moral supervenience, Railton offers the following: "Twins who were genuinely identical in their psychological, physiological, and circumstantial condition (if that were possible) would also be identical in their moral character, and we could not alter this without altering one or the other of them in some nonmoral way—making one more attentive to the interests of others, say, or bolder in defense of his principles. This is the idea that moral properties are not 'self-standing,' they are *supervenient* on nonmoral properties" (Railton 1996: 65).

5. A closely related position is known as non-reductive naturalism, which is a different version of naturalistic moral realism that has been defended by Brink (1989: 9, 156–63) and Sturgeon (1984: 60–63). See (Railton et al. 1992b: 168–74).

6. Railton describes "polywater" in (Railton 1989: 161). Other examples of how a reduction can eliminate or explain away properties and substances are seen in the fates of phlogiston, caloric fluid, and vital force (Railton 1993b: 317).

7. Also see (Railton 1993b: 317, 325).

8 Another area where Shafer-Landau and Railton disagree is over moral explanations. Railton thinks that we often explain people's actions with reference to moral facts, features and properties, so moral explanations are viewed as causally efficacious. Shafer-Landau does not accept causal moral explanations.

9. The issue of how moral facts as natural facts can motivate, brings us back to the topic of internalism (§§2.1, 2.2b, 2.3a). A version of strong internalism says that moral notions provide sufficient motive or reason for acting morally, whereas weak internalism says that moral notions provide only *some* reason or motivating force. Railton adopts a weak version of internalism (1986a: 168, 1986b: 29, 1989: 171). He claims that "what is intrinsically valuable for a person must have a connection with what he would find in some degree compelling or attractive, at least if he were rational and aware. It would be an intolerably alienated conception of someone's good to imagine that it might fail in any such way to engage him" (Railton 1986b: 9).

10. Also see (Railton 1993a: 292).

11. See also (Railton 1986b: 9–17). Many dilemmas can arise due to conflicts between various nonmoral goods. Railton says that his version of moral realism actually emphasizes conflict, not "world-wide consensus" (Railton 1986a: 199). Railton states: "The multiple pulls one feels do not incline one to wonder whether anything real is at issue, rather, they make one especially aware of just how real the competing considerations seem to be and how little we can mitigate them by sheer decision or convention or changing attitudes" (Railton 1992a: 740).

12. Street's (2006) paper offers a much more detailed and elaborate challenge to the moral realist. Here I am only highlighting her main points.

13. Street actually claims that Railton's account is not genuinely realist because he does not assert the independence of evaluative facts from evaluative attitudes strongly or cleanly enough. By her lights, because his "full information" account of nonmoral

good implies that evaluative facts will be different when evaluative attitudes are different, that shows that the evaluative facts are not truly *independent* from the evaluative attitudes (2006: 136–37, 161). I don't have the space to go into it, but I suspect that Railton may address this objection with his notion of "relational goodness" (Railton 1986b: 10). If one overemphasizes the "independence" of values and moral truths, then one ends up with an absolutist version of intrinsic value; Railton though, does not equate intrinsic value with absolute value. And he similarly does not speak of absolute goodness, but only relational goodness (1986b: 9–10). Copp (2008) defends *his* version of naturalistic moral realism against Street's argument.

14. Another way to say it is that the *form* of the moral sense is sentiment, while the *content* is not.

15. With regard to the consensus on right and wrong not depending on religion, Haught would disagree with Dawkins here, because Haught holds that the virtues that we value the most today are eternally grounded, and we owe it to our ancestors who believed this about virtues and are responsible for their being passed down to us (Haught 2000: 122).

Part II

EVOLUTION AND NORMATIVE ETHICS

Chapter 6

Evolution and Virtue Ethics

Here is a famous hypothetical scenario set up by the moral philosopher Bernard Williams. It has provoked much discussion. Williams asks us to imagine that a vicious general, Pedro, is about to execute twenty innocent Indians. Pedro is doing this in order to deter other Indians from rebelling. Pedro offers a chance visitor, Jim, the opportunity to kill one of these Indians himself. If Jim takes up the offer, the general will spare the other nineteen; otherwise, all twenty will die. Should Jim take the offer? (Williams & Smart 1973: 98–99).

I open with this scenario not because it intuitively leads us into virtue ethics, the main topic of this chapter; but because it intuitively leads us into *normative* ethics more generally, which is the main focus of Part II of this book. Normative ethical theories provide advice about what we ethically should do. The ethical theories examined in Part I of this book (error theory, expressivism, moral relativism, and moral realism), on the other hand, do not provide ethical guidance for decision making. Those metaethical theories attempt to answer different questions about ethics. They will not help Jim decide on what he should do when he finds himself in his predicament with Pedro the vicious general. Normative theories, though, *do* focus on very practical ethical questions. This is what separates normative ethical theories from metaethical theories. Each of the normative ethical theories we will examine has a suggestion about what we should do, ethically.

The first normative ethical theory we will consider is virtue ethics. In this chapter I first lay out the main points of virtue ethics as they have been developed both by a classic virtue theorist—Aristotle (337 BCE)—and a contemporary proponent of Aristotelian virtue ethics—Martha Nussbaum (1988).[1] The main ideas of virtue ethics include: virtues as traits of character, habits and excellences, happiness/flourishing, exemplars (role models), and spheres of human experience. Having outlined the main ideas of virtue ethics, I then discuss how some theorists (Michael Ruse, William Casebeer, and Jonathan Haidt & Craig Joseph)[2] have characterized the relationship between virtue ethics and evolution. Lastly, I examine what different models of evolution imply about virtue ethics.

6.1 (ARISTOTELIAN) VIRTUE ETHICS

Let's first lay out the main points of virtue ethics that are pertinent to the present discussion.

In the study of ethics, going all the way back to antiquity, concepts like courage, generosity, honesty, humility, etc., are known as moral virtues. Moral virtues are regarded as a type of behavioral characteristic of individual persons. To be a characteristic or firm disposition indicates something more substantial than simply a once-performed action. Genuinely to be a virtue, what we mean is that the person has a habit of performing particular kinds of acts. If a person possesses the virtue of generosity, for example, that means that he or she has the habit of performing generous acts.

Moral virtues are developed through repeated actions. Humans are born with the potential to develop these virtues, Aristotle claims, not born with fully formed virtues.

> Virtue of character results from habit. . . . Hence it is also clear that none of the virtues of character arises in us naturally. . . . And so the virtues arise in us neither by nature nor against nature. Rather, we are by nature able to acquire them, and we are completed through habit. (Aristotle 337 BCE: 18)

And whether the moral virtue will be developed or not depends on an individual's choice of actions. Aristotle maintains that humans are rational creatures, so humans are aware of what they are doing; they thus have the choice about which actions to perform, and which actions to perform repeatedly. As Aristotle puts it:

> We have found, then, that we wish for the end, and deliberate and decide about things that promote it; hence the actions concerned with things that promote the end are in accord with the decision and are voluntary. The activities of the virtues are concerned with these things [that promote the end]. Hence virtue is also up to us, and so also, in the same way, is vice. (Aristotle 337 BCE: 37)

Humans can reflect on the habits they have, and decide whether they are satisfied with their habits, or resolve to change them (think of turning over a new leaf for the New Year). This kind of developmental process not only applies to moral development, but physical and mental development too.

Think of the repeated actions necessary to achieve physical fitness. In education, think of the challenge of mental exercises performed to enhance one's mental development, or more specifically, think of writing ability, both mental and physical—whether pen or keyboard.

One's own character, then, is one's responsibility, because the actions one repeatedly performs yield the habits/traits that one possesses.[3] And it is not only external actions that are of relevance here, but also internal states of emotions. The virtue of courage is not merely about courageous actions, for example, says Aristotle, but about how one handles the feeling of fear. Is one overcome and paralyzed with fear, or does one lack fear altogether (fearless)? For Aristotle, neither of those indicate the trait of courage; the virtue of courage lies somewhere between those two extremes. Virtues are about feeling and actions, he says, and "these admit of excess, deficiency, and an intermediate condition. We can be afraid, for instance, or be confident, or have appetites, or get angry, or feel pity, and in general have pleasure or pain, both too much and too little, and in both ways not well" (Aristotle 337 BCE: 24). One can think of the various traits we develop in ourselves over time as the building blocks of our character. Our character is based on which actions we repeatedly perform, and how we handle a range of emotions. In fact, because of the centrality of the concept "character," this kind of ethical theory we are discussing is sometimes called a "character ethic" instead of a "virtue ethic."

Although being a habit is a necessary component of being a virtue, it is not a sufficient component. This is because a habit can be considered a *bad* habit. Smoking cigarettes can be a habit, for instance. But because smoking cigarettes is a habit, that in itself doesn't make it virtuous to smoke. By definition, virtues are not *bad* habits; virtues are *good* habits. More than merely good, though; virtues are *excellent* habits. To possess virtues is to possess excellences. As a way to distinguish between traits that are excellences and traits that are vices (bad habits), virtue theorists refer to the concept of happiness/flourishing/well-being (Aristotle wrote in Greek, so the translation of this concept is not perfect here).

One might simply say that a trait is an excellence since it aids that person in achieving happiness and well-being, and contributes to that person's flourishing. Another term Aristotle is fond of using in this context is "the good." With virtues, one can achieve the good life, he argues. Aristotle's point is that if we consider virtuous traits such as courage, temperance, honesty, and friendliness, and imagine a person who has them and a person who lacks them, we can generalize and say that the person *with* those virtues will be the one who flourishes and achieves well-being. As Aristotle puts it: "The belief that the happy person lives well and does well also agrees with our account, since we have virtually said that the end is a sort of living well and doing well" (Aristotle 337 BCE: 10).

Ethics is not an exact science, Aristotle clarifies, and we should only seek "exactness in each area to the extent that the nature of the subject allows" (Aristotle 337 BCE: 2). When discussing the subject matter of ethics, Aristotle will often use qualifiers such as "by and large," and "for the most part." In discussing ethics, he does not presume to be able to achieve the same level of precision that can be achieved in logic or science. So he says, "Since this is our subject and these are our premises, we shall be satisfied to indicate the truth roughly and in outline; since our subject and our premises are things that hold good usually" (Aristotle 337 BCE: 2). Nor does he claim there to be a guarantee that if one possesses these virtues, then one will achieve the good. This is not because the traits are not excellences, he argues, but because other circumstances of one's life, such as misfortune, lack of opportunity, or bad health, may undermine one's well-being, for instance. As he states: "Happiness evidently also needs external goods to be added . . . since we cannot, or cannot easily, do fine actions if we lack the resources" (Aristotle 337 BCE: 11). For Aristotle, friends are considered the "greatest external good" (Aristotle 337 BCE: 148).

As mentioned above, Aristotle describes human beings as rational animals who are aware and responsible for what they choose to do. There is another significant dimension of human beings that Aristotle emphasizes, though: human beings are social animals. By this, Aristotle means that humans live with others in groups and flourish in a social environment (Aristotle 337 BCE: 148). Many of the virtues he discusses are social virtues, such as gentleness and friendliness. We can develop these traits (or their associated vices) in the context of dealing with others. With regard to the context of dealing with others, Thomas Hurka (2001: 21) uses the example of "indifference to others' pain" as indicative of the vices of callousness, sloth, and apathy. Some virtues that would be relevant to this context include compassion and justice, for instance.

A social environment, then, is very important for the development of virtue. Another concept that comes into play here is the notion of a role model. Virtue ethicist Julia Annas goes so far to say that "Aristotle's account of moral development . . . relies on role-models" (Annas 2003: 28). This is because human beings naturally learn from others; they are natural imitators (most of the time they don't even realize it). Because human beings are natural imitators, Aristotle thinks it is very significant for an individual to have good role models in their social environment, most especially in their youth, when foundational habits are forming. For the most part, one is more likely to develop moral virtues if one is exposed to role models who possess moral virtues. We can very easily imagine being exposed to role models who *lack* moral virtues, and given our social nature, we may develop the same bad habits in ourselves. It is possible to understand "social environment" very broadly here, to include cultural environment, which for us would introduce various forms of media such as literature, television, audio, video, etc., as possible ways in which people will be exposed to various role models. So Holmes Rolston III, for instance, has claimed that the Good Samaritan "has been a role model for millennia" (Rolston III 2004: 238). And obviously Rolston means this as an example of a text-based role model.

Instead of the term *role model* some theorists prefer the term *exemplar*. Linda Zagzebski (2010), for example, has worked with virtue ethics and she calls her version of virtue ethics *exemplarism* because she says the focus of her ethical theory is not *virtues* but *virtuous persons*, that is, exemplars. Zagzebski agrees with Aristotle that human beings have a natural "imitating mechanism," and a "wanting to be like." She suggests that as children the very roots of our evaluative judgments stem from thoughts such as "I want to be like that person," or "I don't want to be like that person" (Zagzebski 2010: 52–53).

At the very beginning of this section on virtue ethics, I mentioned that the study of (virtue) ethics goes all the way back to antiquity. The Greeks were not the only ancients to study virtues, however; so the philosopher Bina Gupta argues that the resources of different traditions should be brought to bear in a thorough analysis of virtue ethics. She writes:

> In the last three decades, a number of major moral philosophers have tried to free themselves from the clutches of Kantianism and Utilitarianism by developing an ethics of virtue. However, it is disappointing to note that in their conceptions of virtue ethics, they could not go beyond a revised form of Aristotelian ethics. The large virtue traditions of India and China totally escaped them. (Gupta 2002: 30)

The philosopher Amy Olberding (2012) has answered Gupta's call. Like Zagzebski mentioned above, Olberding prefers the term *exemplarism* over *virtue ethics*. Olberding distinguishes exemplarist theories from virtue ethics theories by saying that exemplarism maintains that the mechanism for the acquisition of virtues falls squarely on the shoulders of exemplars (Olberding 2012: 42). She uses the ancient text *The Analects of Confucius* and makes the case that not only can Confucius' teachings be understood through exemplarist theory, but that Confucius himself serves as an exemplar from which his students can learn about virtues.[4] The cultural roots of different traditions of virtue ethics brings us to the last virtue ethicist we will consider in this section. Martha Nussbaum defends an Aristotelian version of virtue ethics (Nussbaum 1988: 43, 51). She acknowledges, though, that different cultural traditions have different "lists of virtues," or catalogues that are regarded as significant in that particular cultural tradition (Nussbaum 1988: 34). Nussbaum suggests not that we simply accept Aristotle's particular list of virtues, but that we try to see how he created his list, and see what goes into identifying virtues and vices.

Aristotle, she says, held that there are "features of humanness that lie beneath all local traditions and are there to be seen whether or not they are in fact recognized in local traditions" (Nussbaum 1988: 33). She agrees with Aristotle on this, and argues that all human beings share enough features in their common humanity, so that some virtues are necessary and important no matter what particular cultural circumstances we could imagine. She prefers to put the point in terms of *spheres of human experience*. She holds that there are common spheres of human experience that figure in more or less any human life, and these have to do with fear, bodily appetites, management of one's property, living together with others, etc. For each sphere, there are corresponding virtues that are relevant to it, for example, fear (courage), bodily appetites (moderation), management of one's property (generosity), living together with others (truthfulness), etc. (Nussbaum 1988: 35). Virtues and vices always relate to a sphere of experience.

The first part of the virtue theorist's inquiry, she says, is to identify these basic experiences and basic spheres that would have to be dealt with in any human life. To emphasize the very basic and primal nature of these experiences, she sometimes uses the phrase *sphere of the grounding experiences* (Nussbaum 1988: 36–38, 46–48). The second part of the virtue theorist's inquiry is to explain what the appropriate response should be in that sphere of experience, that is, what the virtue precisely consists in (Nussbaum 1988: 38). So even if lists of virtues seem to differ between cultural traditions, Nussbaum argues that often what differ are the precise specifications of the virtues, not the spheres of experiences. Also, though, we *do* find much agreement across cultural traditions of what character traits are virtuous. Nussbaum sees "a great deal of overlap and convergence among cultures at the level of these experiences. We can see ourselves in their accounts of fear and grief and hunger and delight, and they themselves in ours" (Nussbaum 1995: 120).

Having outlined the main dimensions of virtue ethics, we will now move on to discuss how virtue ethics is seen from an evolutionary perspective. But first let me clarify one last point. We may wonder why virtue theory (or exemplarist theory) is regarded as an "ethic." Isn't it merely a theory of behavioral development? Which part of it is the normative ethical part? The answer is that normative ethical theories provide practical advice about what we ethically should do and how we should live. Virtue ethics provides this: it recommends what kind of person we should strive to become, and in doing so recommends what we should do. We should seek good habits, virtues, and this involves doing what a virtuous person would do. And we should avoid bad habits, vices. In Nussbaum's terminology, we should seek what is *appropriate* in the spheres of experiences that we necessarily encounter in living a human life. Not that what is appropriate is always obvious, or that this is an easy task. Think back to the example of Jim and Pedro. What should a virtuous person do in that circumstance? Virtue theorists maintain that the elements of virtue theory remain useful as we deliberate and reflect on how to live a life.[5]

6.2 VIRTUE ETHICS AND EVOLUTION

Virtue ethics has become very popular among moral theorists.[6] Even ancient Aristotelian virtue ethics continues to have defenders. And even though Aristotle does not talk about ethics from an evolutionary perspective, fairly recently virtue ethics has been analyzed from an evolutionary perspective.

The well-respected moral philosopher Bernard Williams, though, has claimed that this "neo-Aristotelian enterprise" might "require us to feign amnesia about natural selection" (Williams 1995). Williams' worry about a neo-Aristotelian project is that such a project works best, he says, "if you can help yourself to Aristotle's cosmology," rather than entertain the possibility that "the evolutionary success of humanity, in its extremely brief period of existence," rests "on a rather ill-assorted *bricolage* of powers and instincts" (Williams 1995: 199). And in case you are unclear what a "bricolage" is, it is a kind of artistic piece created out of a diverse and disjointed range of things that happen to be available.

Williams' challenge to an Aristotelian virtue ethics amounts to an evolutionary challenge, because he suggests that if we believe natural selection has over time shaped and created human beings, then Aristotle's ethical ideals do not apply as well to human behavior. Aristotle's cosmology and his characterization of human nature simply do not fit into the world of evolutionary biology. Williams writes:

> The first and hardest lesson of Darwinism, that there is no such teleology at all, and that there is no orchestral score provided from anywhere according to which human beings have a special part to play, still has to find its way fully into ethical thought. (Williams 1983: 565)

So in this chapter in the context of reviewing how some theorists have characterized the relationship between virtue ethics and evolution, I shall consider whether Williams' evolutionary challenge can be met.

Before I try to answer Williams' charge by drawing on recent theorists who have worked explicitly with virtue ethics and *evolution*, I will briefly touch on Paul Churchland's take on Aristotle's virtue ethics (1996, 1998). Churchland is a philosopher who has used the concept of neural networks to try to explain the workings of the mind. But in a few articles on *ethics*—though he does not explicitly mention evolutionary implications for virtue ethics—he has emphasized that

> the common picture of the Moral Man as one who has acquiesced in a set of explicit rules imposed from the outside—from God, perhaps, or from society—is dubious in the extreme. . . . It is much more accurate to see the moral person as one who has acquired a complex set of subtle and enviable skills: perceptual, cognitive, and behavioral. This was, of course, the view of Aristotle. . . . Moral virtue, as he saw it, was something acquired and refined over a lifetime of social experience, not something swallowed whole from an outside authority. It was a matter of developing a set of largely inarticulable skills, a matter of *practical wisdom*. Aristotle's perspective and the neural network perspective here converge. (Churchland 1996: 105–6)

Like Aristotle, Churchland claims that moral knowledge should be portrayed as "a *set of skills*," a matter of practical wisdom (Churchland 1998: 79). For Churchland, that set of skills is diverse; it includes perceptual, recognitional, and social skills (Churchland 1998: 80, 86). Moral virtues, then, for Churchland are simply

> the various skills of social *perception*, social *reflection*, *imagination*, and *reasoning*, and social *navigation* and *manipulation* that normal social learning produces. (Churchland 1998: 86)

So, for example, Churchland says our ability to grasp patience and courage far outstrips our capacity for verbal definitions of these notions (Churchland 1996: 101). Ethical knowledge is more a matter of *knowing how* rather than *knowing that*.

With this definition of virtue, Churchland then argues that Aristotle's concept of *moral character* is also useful. As Churchland puts it:

> A person's unique moral character is just the individual profile of his perceptual, reflective, and behavioral skills in the social domain. (Churchland 1998: 87)

Churchland views one's character as "a slowly acquired network of skills," and like Aristotle, Churchland argues that as the suite of one's virtues (that cluster of skills) gets developed, that path is the development of one's moral character (Churchland 1998: 87).

Additionally, Churchland upholds the Aristotelian view that all human beings share enough features so that some virtues are necessary and important, no matter what particular cultural circumstances we could imagine (Churchland 1996: 107).

As we discuss evolution and virtue ethics, I think we should bear in mind Churchland's model of virtue ethics, and how it is situated in Churchland's theory of the mind as a collection of neural networks. It may also be interesting to note that when E.O. Wilson has forayed into ethical matters he has critiqued religiously based ethics, but has also sketched an account of ethics that he admits has "roots that go back to Aristotle's *Nichomachean Ethics*" (Wilson 1998: 248).

Turning now to more explicitly evolutionary considerations, philosopher Michael Ruse is an example of an evolutionist who has examined Aristotle's virtue ethics from an evolutionary perspective (Ruse 1991a). Ruse says that in Aristotle "There is much to gladden and encourage the Darwinian" (Ruse 1991a: 76). First, he says, there is Aristotle's "absolutely central recognition of human sociality." Second, there is Aristotle's view that actions are justified in terms of their payoff to us; we are happier or fuller. And third, with respect to many of the Aristotelian virtues, Ruse sees no reason why selection should not have promoted them as fully for women as it has done for men (Ruse 1991a: 80).

Ruse does identify two areas where Aristotelian ethical theory seems to clash with Darwinian thinking, though, and they seem to link with Williams' worry. The first is that Aristotle emphasizes nurture more than the Darwinian, says Ruse. Ruse presumably has in mind that nurture is significant for a person to be able to successfully travel along his or her path toward eudaimonia, which was preset by nature. Though environmental conditions are taken seriously by the Darwinian, Ruse detects that Aristotle simply takes them as more significant for how an individual will turn out. The second (and biggest drawback), says Ruse, is Aristotle's notion that we are essentially rational, that is our purpose and "end." This is the familiar Aristotelian view that our rational nature points to our ultimate purpose. This, too, is that unique purpose of a human person—allegedly part of Aristotelian theory—which is no part of the Darwinian picture.[7] So although Ruse says that in Aristotle there is much to gladden the Darwinian's heart, it is still questionable whether neo-Aristotelian virtue ethics can expunge any trace of Aristotle's ancient cosmology, and thereby meet the evolutionary challenge.

The philosopher William Casebeer (2003) has defended a virtue ethics account in which he claims to bring "Aristotle up to date, biologically speaking" (Casebeer 2003: 47, 49). Casebeer holds that Aristotle is correct in characterizing ethics as "ultimately a practical discipline" (Casebeer 2003: 47). As a virtue theorist, Aristotle views ethics as "a matter of embodying the appropriate states of character so as to function well and achieve *eudaimonia*" (Casebeer 2003: 114). Casebeer notes that the Aristotelian term *eudaimonia* is "variously translated as happiness, success, well-being, and . . . proper functioning," but Casebeer's favorite translation of this concept is "proper functioning" (Casebeer 2003: 43). "Proper functioning" fits into virtue ethics, Casebeer claims, because a virtue ethicist will say that the "person reasoning well will act so as to cultivate those states of being—the virtues—that enable him to function properly" (Casebeer 2003: 43). Virtue ethics is thus a normative ethical theory because it "considers what states of character we ought to cultivate in ourselves" (Casebeer 2003: 12).

By engaging with Aristotle's notion of "proper functioning," Casebeer has formulated a response to the suggestion that Aristotle's account is overly rooted in an ancient cosmological outlook. Casebeer suggests that in the philosophy of biology today there are various analyses of the concept of function. Casebeer prefers the analysis offered by Godfrey-Smith (1994), which frames a function in evolutionary terms. Functions, for Godfrey-Smith and Casebeer are "dispositions and powers which explain the recent maintenance of a trait in a selective context" (Casebeer 2003: 52–53; Godfrey-Smith 1994: 356). Functions "are a result of interaction between organisms and environments" and so are set in place over evolutionary time (Casebeer 2003: 55).

For Casebeer, then, combining this "contemporary biologically oriented notion of function" with "an appropriately naturalized Aristotelian virtue theory" gives us a compelling normative ethic (Casebeer 2003: 56). Casebeer holds that "Aristotle's virtues are in fact virtuous" in that "they help us to live a fully functional life" (Casebeer 2003: 156). Casebeer, therefore, would not find Williams' charge unanswerable, since Casebeer thinks the account of proper function he uses in his neo-Aristotelian virtue ethics fits "perfectly well within a materialist ontological framework," and it is an account that explicitly takes on board an evolutionarily defined sense of "proper function" (Casebeer 2003: 54).

Another attempt at linking Aristotelian virtue theory with evolution comes from Jonathan Haidt and Craig Joseph, two psychology researchers who have published a few articles on virtue ethics (2004, 2007). Haidt and Joseph look at how different cultures, East and West, think about the moral domain. They are interested in identifying the "the psychological primitives that are the building blocks from which cultures create moralities that are unique yet constrained in their variations" (Haidt & Joseph 2007: 381).

Since Haidt and Joseph are interested in cross-cultural dimensions of ethics, they argue for "a broader conception of morality," unlike other predominant conceptions of morality they observe that narrowly focus on harm, rights, and justice (Haidt & Joseph 2007: 367). The broader conception of morality they defend fits best with virtue theory, they say. They claim to have a firm commitment to virtue theory, because they find it to be most psychologically sound (Haidt & Joseph 2007: 387; 2004: 62).

Haidt and Joseph argue that

> the adult mind is full of moral intuitions, which are like little bits of input-output programming connecting the perception of a pattern in the social world (often a virtue or a vice) to an evaluation and, in many cases, a specific moral emotion (e.g., anger, contempt, admiration). (Haidt & Joseph 2007: 379–80)

Many of these moral intuitions, they claim, are "organized, to some extent, in advance of experience," and so this is how "the moral mind is partially structured in advance of experience" (Haidt & Joseph 2007: 367, 381). For Haidt and Joseph, these moral intuitions have a biological and evolutionary origin. So, what to Williams might seem like a bricolage-like assembly of powers and instincts put in place by natural selection, Haidt and Joseph see as amenable to organization into five separate domains: harm/care, fairness/reciprocity, in-group/loyalty, authority/respect, and purity/sanctity. Because they see these domains across cultures, they take them to be "the most important ones for understanding human morality and moral diversity" (Haidt & Joseph 2007: 385).

Like all virtue theorists, they understand virtues to be traits/characteristics of persons that are morally praiseworthy, and, as Churchland emphasizes, social skills.

> To possess a virtue is to have extended and refined one's abilities to perceive morally relevant information so that one is fully responsive to the local socio-moral context. To be kind, for example, is to have a perceptual sensitivity to certain features of situations. (Haidt & Joseph 2007: 386)

Virtue and vice words provide a vocabulary for talking about emotional reactions (Haidt & Joseph 2007: 383).

Another aspect of virtue theory they find to be psychologically sound is the account of how virtues are acquired. Moral intuitions, which are organized in advance of experience, can develop into virtues over time, through exposure to people and stories. Haidt and Joseph see this as a crucial tenet of virtue theory—that the virtues are acquired inductively, through exposure to—sometimes with efforts to copy—many examples of the virtue in practice (Haidt & Joseph 2007: 386). "Only over time will the moral learner recognize what information is important to notice and retain, and what can safely be disregarded," they claim (Haidt & Joseph 2007: 386).[8]

As Haidt and Joseph view things, each culture must deal with adaptive challenges. Because of these adaptive challenges, they view each culture today as having an evolutionary preparedness. By correlating an adaptive challenge (such as protection and care for young, or avoiding microbes and parasites) with a particular domain of adaptive triggers (such as suffering/distress, dealing with waste products), they see how relevant virtues and vices correlate with these challenges and domains (the virtues of caring and kindness versus the vice of cruelty, for instance, with protection and care for the young, and the virtues of temperance, chastity, and cleanliness with avoiding microbes and parasites). The five separate domains they have identified are: harm/care, fairness/reciprocity, in-group/loyalty, authority/respect, and purity/sanctity.

They claim to find "a surprising degree of similarity" in how cultures deal with these five adaptive challenges (Haidt & Joseph 2007: 384). Further, they observe that four of their proposed domains "(all but purity) appear to involve psychological 'building blocks' that are present in other primates" (Haidt & Joseph 2007: 385). This overlap with other primates leads them to their evolution-based theory of the virtues.

Reflecting on the five domains of adaptive challenges, Haidt and Joseph can infer various things about how our ancestors lived:

> They were ultrasocial creatures . . . finely tuned for (1) rearing children and helping kin, (2) selectively cooperating with non-kin while remaining vigilant for cheaters, (3) forming strong in-groups for the purpose of cross-group competition, (4) organizing themselves hierarchically, and (5) attending to each other's physical states, and altering interactions and contacts accordingly. (Haidt & Joseph 2007: 385)

In terms of virtue theory, their model explains how "the content or structure of a virtue is not completely culturally relative" because as "Aristotle pointed out . . . what it means for a personality characteristic to be a *virtue*, and not simply a behavioral regularity, is largely that it consists in functioning well in a specific 'sphere of existence,'" a concept that they think is "similar to what evolutionary biologists would recognize as persistent adaptive challenges and other types of environmental constraint" (Haidt & Joseph 2007: 387). Overall, then, for Haidt and Joseph virtues "are therefore quite at home in a scientific theory of moral functioning based on evolutionary psychology and cultural psychology" (Haidt & Joseph 2007: 387). As psychology researchers, they have attempted to "explain how the intuitive, evolved foundations of morality are elaborated by cultural activity into the complex, diverse moral functioning that mature human beings display" (Haidt & Joseph 2007: 368). So even if Williams is correct that a neo-Aristotelian virtue ethics project works *best* "if you can help yourself to Aristotle's cosmology," rather than entertain the possibility that "the evolutionary success of humanity" rests "on a rather ill-assorted *bricolage* of powers and instincts," such a project still seems to *work very well* even without assuming Aristotle's cosmology and yet fully acknowledging that the assortment of human instincts are evolution based. Casebeer's virtue ethics account can face Williams' challenge because Casebeer defines "functioning well with virtues" in evolutionary terms. And Haidt and Joseph show how we can assemble the range of moral intuitions the moral mind possesses in advance of experience, in a way that shows the appropriateness of a virtue ethics for creatures like us who are the product of biological evolution.

6.3 MODELS OF EVOLUTION AND VIRTUE ETHICS

6.3a Darwin and Virtue Ethics

Turning now to Darwin's own perspective, let us see how the components of a virtue ethics might fit with Darwin's model of evolution.

As Ruse had pointed out, both Aristotle and Darwin emphasize humans as social creatures. Darwin holds that "every one will admit that man is a social being" (Darwin 1871: 84). Because of that fact about human nature, Darwin says that certain virtues must be practiced to associate in a social body, and they are the same virtues we still recognize today, he thinks (Darwin 1871: 61). Darwin is referring to what virtue theorists would call "social virtues."

Darwin frequently refers to virtues. He identifies the virtues courage, humanity, kindness, fidelity, truth, honesty, prudence, temperance, and others (Darwin 1871: 61–63). He singles out courage as a quality that "has universally been placed in the highest rank" (Darwin 1871: 95). Because humans are social creatures, thinks Darwin, they would inherit the tendency to be committed to friends and obedient to leaders, for these traits are common for social animals (Darwin 1871: 84). He even thinks that virtuous tendencies can be inherited (Darwin 1871: 102).[9]

Imitation is another aspect of the social nature of human beings that is relevant to virtue ethics. Aristotle points out that imitation is a natural way that humans learn. With the help of role models we learn which virtues are important. Darwin, too, recognizes the importance of imitation (Darwin 1871: 81). Darwin even uses the term "principle of imitation" and he says imitation comes very naturally to social creatures and imitation has many benefits (Darwin 1871: 161).

Darwin also insists, like virtue theorists, that habits play an important part in guiding conduct (Darwin 1871: 42, 53). According to him, "The social instinct, together with sympathy, is, like any other instinct, greatly strengthened by habit" (Darwin 1871: 73). And he acknowledges how repeated actions become second nature to us (Darwin 1871: 88). This is an indication of how Darwin disagrees with what we calls a "formal" view of morality, one in which an action is regarded as moral after a victory over opposing desires, but not when it is performed impulsively (Darwin 1871: 87).

Aristotle describes humans as rational animals that are aware of what they are doing and have the choice about which actions to perform, and Aristotle also acknowledges that human beings are capable of a wide variety of emotions. A rational being will flourish only when the agent controls the emotion and not vice versa. Like Aristotle, Darwin refers to this capacity in human beings; he calls it self-command. Darwin says that self-command, self-sacrifice, and the power of endurance are important qualities necessary for the practice of the virtues, which are necessary for the welfare of one's tribe (Darwin 1871: 96).

Overall, Darwin's approach to evolution and ethics fits in rather well with the basics of virtue ethics—virtues and habits that guide conduct and feelings, rooted in instincts yet guided by rational judgment (powers of reasoning), and the influence of role models—all of the virtue ethics elements are there. So even for Darwin, virtue ethics and an evolutionary perspective are not antagonistic, but are complementary.

Let's now look at a more recent model of evolution and how it might work with virtue ethics.

6.3b Dawkins and Virtue Ethics

Even though Dawkins defends genic selection instead of Darwin's individual selection, virtue ethics and a genetic evolutionary perspective still seem to be complementary.

In a book chapter that Dawkins has written on the roots of morality, Dawkins endorses science writer Matt Ridley's *The Origins of Virtue* (1996) as providing "a lucid account of the whole field of Darwinian morality" (Dawkins 2006a: 218). Ridley adopts Dawkins' gene's eye perspective and applies it to human instincts. Ridley claims to be an advocate in the "selfish-gene revolution" (Ridley 1996: 20). From the gene's eye perspective, animals and plants are designed by natural selection to do things for their genes, including their behavior, argues Ridley (Ridley 1996: 18). Yet Ridley can still agree with Aristotle and Darwin that humans are extremely social animals (Ridley 1996: 5). In fact, he claims, humans are more dependent on each other than apes and monkeys are (Ridley 1996: 6). Society evolved as part of our nature (Ridley 1996: 6).

Specifically about virtues, Ridley argues that there is an innate potential for virtues such as generosity and trustworthiness (Ridley 1996: 123). There is no claim here that we are born with these virtues. The claim is that we have the potential for them. This is a point that is central to Aristotle's virtue ethics. And just as Aristotle claims that a virtuous life tends to be a life of flourishing, for Ridley, "emotional habits pay. The more you behave in selfless and generous ways the more you can reap the benefits of cooperative endeavour from society" (Ridley 1996: 141). The way that Ruse put the point, recall, is that in a Darwinian and Aristotelian ethic we are justified in doing what we do because of the payoff, we are happier or fuller, says Ruse.

About selfish genes and ethics from an evolutionary perspective, Ridley's main point is this:

> Our minds have been built by selfish genes, but they have been built to be social, trustworthy, and cooperative. That is the paradox this book has tried to explain. Human beings have social instincts. (Ridley 1996: 249)

Dawkins agrees with Ridley on this. In Dawkins' preface of the thirtieth anniversary edition (Dawkins 2006b) of *The Selfish Gene*, Dawkins says that his 1976 book could have been called "the cooperative gene" since genes can "cooperate" even though they are "selfish." This would have saved a lot of confusion that arose over the years due to the title, Dawkins relates.

> Another good alternative to *The Selfish Gene* would have been *The Cooperative Gene*. It sounds paradoxically opposite, but a central part of the book argues for a form of cooperation among self-interested genes. (Dawkins 2006b: ix)

Since humans naturally have social instincts that have been genetically put in place, what Darwin calls imitation and virtue theorists call role models also has a natural fit with a genic selectionist model. Dawkins himself maintains that humans "undoubtedly have a strong tendency to learn from and copy admired role models" (Dawkins 2005). We noted earlier in §6.1 that role models need not be made of flesh and blood only, but could also be text based. An interesting variation on this comes from the perspective of the naturalist philosopher Daniel Dennett who defends a genic selectionism like Dawkins'. Even though as a naturalist Dennett makes the case that God does not exist, yet given our social predisposition to follow role models (flesh and blood or not), Dennett acknowledges the possibility of viewing God not only as a being "watching you" but as a being that humans can try to emulate. Dennett writes:

> The God who is watching you need not be seen to be either list-making Santa or Orwell's Big Brother, but instead a hero or "role model," as we say today, someone to emulate rather than fear. If God is just, and merciful, and forgiving, and loving, and the most wonderful Being imaginable, then anyone who loves God should want to be just, and merciful, and forgiving and loving. (Dennett 2006: 283)

Because humans have social instincts that are genetically based, it does not follow that humans will necessarily and mindlessly follow role models and develop virtues (or vices). For Darwin and Dawkins, humans are rational; they have "self-command," as Darwin calls it. Dawkins too says we can control ourselves even though we have been created by genes: "We, that is our brains, are separate and independent enough from our genes to rebel against them. As already noted, we do so in a small way every time we use contraception" (Dawkins 1989: 332, 271).

Overall, a genetic basis for virtues, sociality, role modeling, and self-command put virtue ethics on a solid evolutionary grounding.

6.3c Gould and Virtue Ethics

As we have seen in previous chapters, Gould agrees that the human genetic constitution does indeed predispose humans to certain behaviors, yet he holds that genetic influence is weaker than cultural influence. Cultural influence is, of course, social influence. The principle of imitation that Darwin discusses is a specific mechanism of how humans become socially and culturally influenced to develop particular traits: as naturally social creatures, humans are great imitators. Another point that Gould should agree with regarding Aristotle's account is that ethics is not an exact science. As mentioned in other chapters, Gould is of the mind that ethical approaches that take "absolutist" ethical stances do not fit well in a "complex and sloppy" world like ours (Gould 1993: 50). Gould's position would seem to fit well, then, with Aristotle's position that ethics is not a subject matter that admits of a high level of precision, but can only yield conclusions that hold for the most part. Part of the reason for this, Aristotle might say, is the high level of contingency involved. For even if one develops and then possesses many virtues, one may end up in an unfortunate state due to many possible external factors.[10] Because of Gould's evolutionary background, he does have deep appreciation for contingencies. Multiplicities of contingencies play a role in the character one develops, and consequently in the diversity of actions, behaviors, and personalities that we observe filling the social world. It would seem that Gould's evolutionary model could easily employ Haidt and Joseph's evolution-based theory of the virtues to help make sense of this diversity.

Gould from time to time will discuss ethical concepts, principles, theories, and traditions, and even though it seems that Gould's model of evolution works well with virtue ethics, Gould himself does not seem familiar enough with virtue ethics to see this. Gould is familiar with some of Aristotle's views, of course, but it does not appear that he has a detailed understanding of Aristotle's ethics. For example, Gould briefly discusses "Aristotle's golden mean between extremes" (Gould 1999: 210). Gould refers to Aristotle's concept of "a golden mean" as a "principle of intellectual life" that says to avoid extremes. But a study of Aristotle's ethical theory shows the golden mean not to be a "principle of intellectual life," but to be nothing other than a way of describing what a virtue is. In §6.1 we saw how Aristotle views feelings and actions as "admitting of excess, deficiency, and an intermediate condition." The "intermediate condition" is also called by Aristotle "the golden mean." And the reason why Aristotle regards it as "golden" is not because it is a "safe" middle ground, but because for many traits it is the mean that achieves the excellence of the sphere of experience in question. We could supplement our understanding of courage as a virtue, for example, by seeing courage as a mean between cowardice and rashness. We could supplement our understanding of friendliness as a virtue by seeing friendliness as a mean between unfriendliness and overfriendliness.

6.3d Haught and Virtue Ethics

What we have seen of Haught's ethical approach sits firmly within virtue ethics. Haught defends a traditional virtue ethics, but he thinks his version of virtue ethics has a special appeal when situated in his model of theistic evolution.

For Haught, traditional virtues like justice, moderation, and gratitude, for example, should be sought after and developed. But unlike the versions of virtue ethics above, for Haught the roots of a genuine virtue ethics are in God and the religious traditions inspired by God.[11] Haught endorses "the same virtues as those prescribed almost universally by the world's religious traditions" (Haught 2000: 133).[12]

Historically, Haught believes that the virtues that we value the most today are eternally grounded, and we owe it to our ancestors who believed this about virtues and are responsible for their being passed down to us (Haught 2000: 122).

A critic of naturalistic ethics, Haught will also be a critic of any naturalistic virtue ethics that tries to account for the virtues we see in humans as an outgrowth of their social nature and that views virtues as excellent ways that humans have learned to handle adaptive challenges in the spheres of existence in which they find themselves, generation after generation.

Part of Haught's argument is that naturalistic ethical theories cannot make sense of those "heroic, prophetic, individuals" who are "attracted to something, having been grasped by something imperishably and absolutely good" (Haught 2006: 162; 2010: 121).

Haught's evolutionary model takes a broad view, so a personal yearning for the good is writ large, cosmologically, and with his cosmological narrative and aesthetic principles Haught wants to draw our attention to this aspect of reality. Specifically about virtue, he says, "The life of virtue contributes to an intensification of the harmony of contrasts in the cosmic process, and this harmony of contrasts is the defining mark of beauty" (Haught 2000: 134–35). This is Haught's explanation for moral progress (§3.3d).

Additionally, even though he says an ethic from an evolutionary perspective will "pursue essentially the same virtues as those prescribed almost universally by the world's religious traditions," when a theistic evolution takes seriously the cosmological aesthetic principle, then

> this pursuit of virtue would now be fired by a much stronger sense that our moral lives are contributing to the ongoing creation of a universe. (Haught 2000: 133)

This is significant because our actions "'would matter' in a way that the acosmic ethic of detachment from nature, proposed by much traditional religion, could not possibly have envisaged" (Haught 2000: 133).

For Haught, pursuing the virtues is part of the good life, and in the light of the cosmological aesthetic principle we have even more reason to practice virtue, because we can see virtue as contributing to "the creative enterprise of intensifying cosmic beauty" (Haught 2000: 134).

6.4 CONCLUSION

We began this chapter by considering what a normative ethical theory is. We then looked at virtue ethics as a normative ethical approach with ancient roots that still enjoys contemporary defenders. In the course of examining how some recent theorists have characterized the relationship between virtue ethics and evolution, we have uncovered several ways to deflect Williams' objection that neo-Aristotelian virtue ethics depends on an Aristotelian cosmology. In viewing virtue ethics from the perspective of Darwin's model of evolution we saw how the main elements of virtue ethics actually appear in Darwin's own treatment of ethics. Even though Dawkins himself did not view virtue ethics from his genic selection perspective, Matt Ridley, whose work Dawkins endorses, shows how a genic selectionist account of evolution can fit with a virtue ethics. Gould's model, too, although Gould himself does not apply his model of evolution to virtue ethics, does seem to work with a virtue ethics approach. Finally, Haught deliberately combines his model of evolution with a virtue ethics approach. And Haught even takes virtue ethics into new territory, he thinks, with the aid of the two principles of his model of evolution, the cosmological narrative principle and the cosmological aesthetic principle.

Haught's virtue ethics is particularly theistic and supports the virtues of faith, hope, patience, and care:

> The fact that the cosmos is even now perhaps in the early phases of its full emergence helps us understand why, religiously speaking, we remain always

> somewhat in the dark, why our answers to the biggest of our questions will always be frustratingly opaque, why we must walk by faith rather than by sight, but also why it makes so much more sense to hope than to yield to despair . . . patience and waiting may be the profoundest expressions of care. (Haught 2003: 142–43)

In the next chapter we turn to natural law ethics, another normative ethical theory that also has ancient roots, but gets more developed in the medieval period of history; and we will examine it from an evolutionary perspective.

NOTES

1. There are many kinds of virtue ethics approaches, but Aristotle's is an example of an early one that is well worked out and has been very influential. Here is what the prominent virtue theorist Alasdair MacIntyre has to say about Aristotle's ethical framework: "If some particular moral scheme has successfully transcended the limitations of its predecessors and in so doing provided the best means available for understanding those predecessors to date *and* has then confronted successive challenges from a number of rival points of view, but in each case has been able to modify itself in the ways required to incorporate the strengths of those points of view while avoiding their weaknesses and limitations *and* has provided the best explanation so far of those weaknesses and limitations, then we have the best possible reason to have confidence that future challenges will also be met successfully, that the principles which define the core of a moral scheme are enduring principles. And just this is the achievement that I ascribe to Aristotle's fundamental moral scheme" (MacIntyre 1984: 270).

2. Another recent defense of an Aristotelian evolutionary virtue ethics is John Lemos (2008). Lemos calls his view "Aristotelian Evolutionary Ethics." Unfortunately I do not have space to discuss it here.

3. On this topic, I recommend Jonathan Jacobs' *Choosing Character: Responsibility for Virtue & Vice* (2001).

4. Donald J. Munro (2005) has explored Confucian virtue ethics alongside evolutionary biology and argues that Chinese virtue ethics is consistent with contemporary evolutionary biology.

5. Rosalind Hursthouse is a virtue theorist who elucidates how virtue ethics is a normative ethic (1996). She applies virtue ethics to the question of abortion, for example. See (Hursthouse 1991).

6. A recent exposition of eighteen different virtues is Comte-Sponville (2001). For a collection of articles about virtue ethics, see Crisp (1996), Crisp & Slote (1997), and Gardiner (2005).

7. I myself prefer Stan van Hooft's interpretation of Aristotle's discussion of our rational nature as more about a stage in one's life, rather than a general description of our essentially rational nature as our ultimate purpose. van Hooft (2006) thinks Aristotle meant that the life of contemplation is the ultimate purpose only for people

at the end of life, those in retirement. If we *only* think of ourselves as rational beings, then the good life would be the life of contemplation and thinking. But this overlooks the other sides of the human person—because, even by Aristotle's own admission, we are not only rational beings, we are also emotional, social, and political beings. I therefore regard the issue of whether Aristotle necessarily links one's telos simply with one's rational nature as controversial.

8. Some psychologists use the term "moral elevation" to describe how people can begin to act virtuously when exposed to others' virtue. See Brooks (2014).

9. Richerson & Boyd (2004: 58, 61) warn about seeing "inherited habits" that Darwin discusses as "biologically" inherited habits.

10. Aristotle admits of what today some theorists call "moral luck." See Nagel (1976) and Williams (1976).

11. Zagzebski (1998: 539) is another example of a theorist who develops virtue ethics with a theological foundation.

12. Haught, in discussing the possibility of extraterrestrial intelligence, even says that "[o]thers out there probably need courage, too" (Haught 2003: 185).

Chapter 7

Evolution and Natural Law Ethics

Today most people react negatively when asked about the ethics of cloning. Cloning advocates, though, say cloning is similar to twinning. One reproductive cloning technique (cloning-to-produce-children) would take the nucleus of an adult cell and transfer it into a human egg, which would grow into an embryo, then a baby, who would then be the genetic twin of the adult donor.[1] The cloned child would not be fully identical to the donor, just in the same way that twins are not fully identical to each other: other factors come into play to give rise to slight differences. So isn't cloning just another kind of reproduction, but with the benefit that it does not leave the genetic profile of a child up to chance? If we don't react negatively to twins, cloning advocates say, we shouldn't react negatively to cloning either. Relatedly, many people react negatively toward the use of performance-enhancing drugs in sports. They say that rather than naturally working hard to enhance one's performance, athletes who use these substances are unnaturally enhancing their performance. Further, if we combine the possibility of cloning-to-produce-children with the desire to produce exceptional athletes, this brings us to the prospect of genetically modified humans, or "gene editing."[2] Parents may desire genetically enhanced children with exceptional athletic, cognitive, or even musical abilities. Critics, though, say that even if scientific techniques might make it possible to genetically modify one's offspring so they have accentuated abilities, parents should not use these techniques, partly because they are unnatural.

One last example is genetically modified foods. Advocates of genetically modified foods point to the benefits of genetically engineered fruits, vegetables, and maybe even livestock.[3] Many people object to genetically modified foods, though, as they do to cloning, because these procedures are unsafe and too unnatural. Foods with all natural ingredients are commonly regarded as the best.

Those who appeal to natural methods of food production and natural reproduction may well be looking at things from the perspective of natural law, an ethical concept of ancient origin. In terms of time frame, just as virtue ethics is a contemporary normative ethical theory with ancient roots, so too does natural law ethics have both contemporary and ancient proponents. In the ancient Greek play *Antigone* (441 BCE), for example, written by Sophocles, Creon the king of Thebes gives the order that the soldier Polyneices who died in battle, was actually a traitor and therefore shall not receive a proper burial. King Creon decrees that Polyneices' body shall be left in a field. Polyneices' sister, Antigone, refused to abide by the king's law, and buries her brother, deliberately defying the king's order. Antigone, when brought before the king to answer for her crime, tells Creon that there is a law much higher than human laws, and she felt duty bound to uphold that law by giving her brother a proper burial.

Aristotle, too, about a century after Sophocles, acknowledges a natural law. As Aristotle puts it:

> Particular law is that which each community lays down and applies to its own members: this is partly written and partly unwritten. Universal law is the law of nature. For there really is, as every one to some extent divines, a natural justice and injustice that is binding on all men, even on those who have no association or covenant with each other. It is this that Sophocles' Antigone clearly means when she says that the burial of Polyneices was a just act in spite of the prohibition: she means that it was just by nature. (Aristotle 350 BCE: Bk I, Chapter 12: 1373b2–10)

The notion of natural law was also carried forward into ancient Rome. Cicero, for example, as a statesman, writes about natural law and how it is foundational to his general understanding of what laws are.

> True law is right reason in agreement with nature: it is of universal application, unchanging and everlasting; it summons to duty by its commands, and averts from wrongdoing by its prohibitions. . . . It is a sin to try to alter this law, nor is it allowable to attempt to repeal any part of it, and it is impossible to abolish it entirely. We cannot be free from its obligations by senate or people, and we need not look outside ourselves for an expounder or interpreter of it. And there will not be different laws at Rome and at Athens, or different laws now and in the future, but one eternal and unchangeable law will be valid for all nations and all times, and there will be one master and ruler, that is, God, over us all, for he is the author of this law, and its promulgator, and its enforcing judge. Whoever is disobedient is fleeing from himself and denying his human nature. . . . Out of all the material of the philosophers' discussions, surely there comes nothing more valuable than the full realization that we are born for Justice, and that right is based, not upon men's opinions, but upon Nature. (Cicero 52 BCE, Bk 3: 152)

Natural law ideas continued to get refined and elaborated in the medieval and modern periods, up until the present day. In this chapter I first sketch out the most influential version of natural law ethics, as it was laid out by its medieval proponent Thomas Aquinas (1225–1274). In chapter 1 I had mentioned that natural law ethics is a normative ethical theory that provides guidance about what we ought to do; it basically recommends that we ask ourselves whether we are following natural law (§1.5). The key natural law elements include how natural law differs from other kinds of laws; how ethics depends on human nature and human natural inclinations; the concept of human goods; and the ethical directive to value human goods. The chapter then reviews how some theorists (E.O. Wilson, Michael Ruse, Craig Boyd, Stephen J. Pope, and Larry Arnhart) have characterized the relationship between natural law ethics and evolution. Lastly, the chapter views natural law ethics from the perspective of our four models of evolution.

7.1 NATURAL LAW ETHICS

In formulating natural law ethics, Aquinas is known for drawing on many sources. He is a medieval Catholic theologian and philosopher who draws from ancient philosophy, from the Bible, and from earlier thinkers in the Christian tradition. Aquinas' background theism substantially influences his version of natural law ethics.[4]

Aristotle's influence is also apparent in many places of Aquinas' work. Chapter 6 described Aristotle's virtue ethics. Aquinas adopts most of what we have seen of Aristotle's virtue ethics, including the notion of natural justice (as quoted in §7.0). Aquinas adds Christian elements to Aristotle's ethical approach: for example, the Christian belief in supernatural happiness, as well as specific references to different kinds of law—distinctions necessitated due to Aquinas' worldview, as we will see. And Aquinas adds particular Christian virtues. The Christian list of virtues that Aquinas promotes differs from Aristotle's list. Aquinas' way of framing natural law ethics as a theistic ethic shares similarities to Cicero's understanding of natural law, encapsulated in the quotation in §7.0. With regard to the Cicero quotation above, Aquinas agrees with Cicero's description of what the natural law is, the source of the natural law, and how the natural law bears on human conduct. As Cicero puts it, right is based on nature, not opinion. Natural law is also about right reason, understood as reason in accord with nature. For Aquinas, too, through the use of human reason we come to know the natural law and what the natural law requires. For Aquinas, "The precepts of the natural law are related to the practical reason" (Aquinas 1270: Qu 94).[5] Further, natural law is seen as having a universal application; it differs from merely local laws. The natural law also has unchanging and eternal dimensions; thus, it cannot be abolished, agree Cicero and Aquinas. Says Aquinas, "As far as its first principles are concerned, the natural law is altogether unchangeable" (Aquinas 1270: Qu 94). The natural law being eternal hints at its source: an eternal being—God.

Aquinas describes, though, how the natural law can be distinguished from the eternal law. For Aquinas, the natural law is an ethical directive, since (by the use of human reason) this natural law directs us as to which actions to perform and which to avoid. The natural law, then, has practical implications and it illuminates our ethical duties and obligations. The way that Aquinas expresses it is that "all the things that are to be done or evils to be avoided belong to the precepts of the natural law which the practical reason naturally apprehends as human goods" (Aquinas 1270: Qu 94). One last point from Cicero is that whoever goes against this natural law defies human nature. Aquinas, too, will regard actions that go against natural law as unethical for the reason that they go against nature.

In Aquinas' view, our human nature was intentionally created and designed by God, and our lives only reach their natural end when they take us closer to God.[6] Aquinas' worldview thus includes the beliefs that God is the creator, and that everything that exists, including human nature, has its ultimate source in God. Aquinas (1270: Qu 90) quotes St. Paul, who, in his *Letter to the Romans* (2: 14–15), wrote that "though the pagans do not have God's law, nevertheless they know the difference between right and wrong for they have the law written on their hearts."[7] Aquinas refers to an even more ancient reference to the natural law in the *Psalms* (4: 6) of the Old Testament of the Bible. There, Aquinas believes that the distinction between eternal law and natural law is made:

> "The light of Thy countenance, O Lord, is signed upon us": thus implying that the light of natural reason, whereby we discern what is good and what is evil, which is the function of the natural law, is nothing else than an imprint on us of the Divine light. It is therefore evident that the natural law is nothing else than the rational creature's participation of the eternal law. (Aquinas 1270: Qu 91)

Aquinas argues that there are different kinds of law: eternal, natural, human, and divine. Aquinas says that the word law (*lex*) is derived from *ligare*, to bind, because it binds one to act (Aquinas 1270: Qu 90). Aquinas understands eternal law as God's plan, which humans cannot fully comprehend. Here is an example of how Aquinas argues for the existence of an eternal law:

> Assuming that the world is governed by divine providence . . . it is evident that the whole community of the universe is governed by the divine reason. . . . And since the divine reason's conception of things is not subject to time but is eternal, this kind of law must be called the eternal law. (Aquinas 1270: Qu 91)

The eternal law is the rational plan of divine wisdom (Aquinas 1270: Qu 93). But the natural law, by contrast, is the rational creature's understanding of the eternal law. The natural law is only a partial glimpse into God's plan for human beings, yet for Aquinas, it is a reliable guide for determining the basic outlines of an ethical life.

> Rational creatures . . . participate in eternal reason in that they have a natural inclination to the proper actions and ends. Such participation in the eternal law by rational creatures is called the natural law. (Aquinas 1270: Qu 91)

Human laws are different because they are simply those laws created by human beings, such as the law that says we shouldn't text message while driving. Divine law, finally, is understood as God's word as contained in the Bible. Aquinas says that in addition to the natural law and human law it was also necessary to have a divine law to direct human action (Aquinas 1270: Qu 91). This is for many reasons, argues Aquinas, one of which is that humans are pointed to the goal of eternal bliss, which exceeds our natural capacities, so divinely revealed law is necessary (Aquinas 1270: Qu 91).

From here on, I will only focus on the natural law, since it is the central ethical component of natural law ethics. The main foundational ethical principle of natural law ethics spells out in more detail what the natural law ethically requires. Aquinas writes that

> all the things to which man has a natural inclination are naturally apprehended by the reason as good and therefore objects to be pursued, and their opposites as evils to be avoided. Therefore the order of the precepts of the natural law follows the order of our natural inclinations. (Aquinas 1270: Qu 94)

According to Aquinas' theory, then, looking at human natural inclinations (human nature) can help us to discern what the natural law ethically requires of us. Aquinas argues that moral standards come to be known through human reason's reflection on the following central natural inclinations of human beings: preserving one's life, propagation, sociability, and knowledge.

Here is how Aquinas outlines these natural inclinations.

> There is in man, first, an inclination to the good that he shares by nature with all substances, since every substance seeks to preserve itself according to its own nature. Corresponding to this inclination the natural law contains those things that preserve human life and prevent its destruction. Secondly, there is in man an inclination to certain more specific ends in accordance with the nature that he shares with other animals. In accordance with this, the natural law is said to contain "what nature has taught all animals," such as the union of man and woman, the education of children, etc. Thirdly, there is in man a natural inclination to the good of the rational nature which is his alone. Thus man has a natural inclination

> to know the truth about God and to live in society. Thus the things that pertain to inclinations of this kind belong to the natural law, such as that man should avoid ignorance, that he should not offend others with whom he must associate, and other related actions. (Aquinas 1270: Qu 94)

If we observe and reflect on human natural inclinations, Aquinas believes, we will recognize that humans are naturally directed toward fundamental values/goods, and all the things to which humans are inclined by nature belong to the natural law (Aquinas 1270: Qu 94). The things to which human beings have natural inclinations are naturally apprehended by reason as good, he says, and therefore are objects to be pursued, while their opposites, as evils, are to be avoided.

In terms of its practical aspect, the ethical directive of natural law ethics is to promote (rather than destroy) these fundamental values. The main ethical principle of natural law ethics says to preserve fundamental human goods; thus, we ought to preserve ourselves, procreate, be sociable, and seek knowledge.

Given the fact that we are rational beings, we have the capacity to realize that not only do we have these natural inclinations, but other human beings have them too. Thus, we ought not to stand in the way of others as they pursue their fundamental goods, (their own self-preservation, etc.). Natural law ethics thus incorporates the golden rule. Let us consider how Aquinas understands how particular ethical directives derive from the natural law. First off, he thinks that some human actions require very little reflection to determine whether to approve or reject them; yet others involve circumstances that not everyone is capable of reflecting clearly about (Aquinas 1270: Qu 100).[8]

"Do not kill," for instance, is derived from the natural law, he argues. He claims it simply is a conclusion that can be drawn from the principle, "Do not do evil to anyone" (Aquinas 1270: Qu 95). Other examples of actions that are contrary to natural law are the killing of the innocent, adultery, and theft (Aquinas 1270: Qu 94), suicide (1270: II–II Qu 64), and lying. "A lie is evil in itself," he argues, for "it is both unnatural and improper for someone to signify in words what he does not have in his mind . . . it is disordered in itself" (Aquinas 1270: II–II Qu 110).[9]

7.2 NATURAL LAW ETHICS AND EVOLUTION

The biologist E.O. Wilson, whom we have already seen (§§3.3b, 4.2, 5.3b, 6.2), has occasionally ventured into ethical theory, and acknowledges that the extensive natural law tradition has been very influential in the Western world and has even influenced the American founders and later reformers

(Wilson 1998: 239).[10] "So compelling are such fruits of natural law theory," says Wilson; yet "to its noble successes must be added appalling failures" (Wilson 1998: 239). Natural law ethics has been used to justify slavery and genocide, claims Wilson.

In an earlier work of his, Wilson argued that natural law ethics is simply "in error" (Wilson 1978: 141). He says natural law ethics is "based on the idea that immutable mandates are placed by God in human nature," yet since the "laws it addresses are biological, were written by natural selection," these biological laws "have been erroneously interpreted by theologians writing in ignorance of biology" (Wilson 1978: 141).

Michael Ruse, a some-time coauthor with Wilson, emphasizes how morality is part of human nature; he claims that "natural selection has made us in such a way that we enjoy things which are biologically good for us and dislike things which are biologically bad for us" (Ruse 1998a: 236). Ruse is familiar with Aquinas' natural law ethics (Ruse 2001: 161–63). He concedes that "Christian morality is natural, in the deep sense that it stems from the way that we are" (Ruse 2001: 185) and Ruse says Darwinians are "sympathetic to the claim that what is natural is what we value" (Ruse 2001: 197).

However, "One cannot simply equate the natural with the good," Ruse explains (Ruse 2001: 197). And he sees it as a flaw in traditional natural law ethics that it tries to equate the natural with the good. A Darwinian *can* be a Christian, though, according to Ruse (2001: 217). But only "a natural-law type position" can be defended on Darwinian grounds, not a traditional natural law ethic.

Stephen J. Pope is a theologian who also perceives there to be a tenuous relationship between evolution and traditional natural law ethics. He says that "no simple synthesis of Thomas's ethics with evolutionary theory is possible. No simple and direct one-to-one correspondences exist" (Pope 1994: 77). He does see value in bringing these two perspectives together, though. The best we can hope for, says Pope, are some "functional equivalences." For example, just as Aquinas sees that nature has a natural ordering, the same goes for contemporary biology (Pope 1994: 78). "Evolutionary theorists," Pope says, "view 'human nature' itself as ordered biologically, as dependent on and participating in the natural world" (Pope 1994: 78). Aquinas also held that humans "are ordered by nature to live with one another and to participate in political community. . . . Thomas recognized the importance of reciprocity in human social life" (Pope 1994: 79, 85). This comes close to what contemporary evolutionists call "sociality," says Pope, a "functional equivalent" to what evolutionists call "reciprocal altruism" (1994: 79). Also, "[e]volutionary theory claims, as Thomas did, that self-love is based in human nature. . . . Nature teaches all beings to seek their own good; even things completely lacking in intellect tend by nature to procure the good for themselves" (Pope 1994: 81).

The philosopher Craig Boyd is a defender of natural law ethics (2007). Boyd also argues that natural law ethics might plausibly incorporate an evolutionary perspective. He draws on Aquinas's version of natural law ethics, and attempts to show how an evolutionary approach is consistent with natural law ethics and how natural law ethics can incorporate recent biological research (Boyd 2004: 226, 233). Boyd uses the example of the prohibition on murder to make his case, since it is easy to see how an ethical prohibition on murder evolved. The prohibition could have arisen due to kin selection or reciprocal altruism (2004: 229). Aquinas views the prohibition on murder as following from natural law.[11]

The political scientist Larry Arnhart has done the most to defend the view that a Darwinian view of human nature can support the natural law reasoning of Aquinas (Arnhart 1998; 2001; 2004a; 2004b). At the very least, Arnhart sees a compatibility between natural law ethics and an evolutionary perspective. One reason is that Aquinas "recognizes the continuity between human beings and other intelligent animals" (Arnhart 2001: 5).

Further, though, Arnhart claims that "biological reasoning about human nature can strengthen the case for natural law by giving it the support of Darwinian science" (Arnhart 2001: 2). In making his case, Arnhart specifically answers Wilson's objection that natural law thinking is premised on mistaken biology. Against Wilson, Arnhart finds many examples in Aquinas' thought of how Aquinas' approach fits with many tenets of contemporary biology. Arnhart uses the examples of monogamous marriage and incest avoidance, for instance, in a way similar to how Boyd discusses the prohibition on murder. About Aquinas, Arnhart says, "The biological character of Aquinas's reasoning about natural law as rooted in natural desires is clear in his account of marriage and familial bonding" (Arnhart 2001: 5). Promiscuity and adultery are contrary to the natural law, reasons Aquinas. Raising a child requires care of a mother and a father, Aquinas says, and promiscuity does not achieve this. Also, men need to be certain about their offspring, says Aquinas; and once again, promiscuity does not achieve this. He considers the commitment to one woman as part of the natural law (Aquinas 1270: Qu 154).

Thus Arnhart says that for Aquinas, "The primary natural end of marriage is to secure the parental care of children; the secondary natural end is to secure the conjugal bonding of male and female for a sexual division of labor in the household" (Arnhart 2001: 5). Arnhart sees this account as perfectly in keeping with a biological approach, and Wilson himself says that in evolutionary terms "sexual activity between males and females served as the principal device for cementing the pair bond" (Wilson 1978: 140–41).

Stephen Pope uses a similar example. Referring to the passage above where Aquinas claims that "there is a natural human concern among men to be certain about their offspring," Pope says that "Thomas . . . maintained that the priority of parental love is grounded in knowledge of paternity" (Pope 1994: 89). "Evolutionary theory again," writes Pope, "provides biological information functionally equivalent to Thomas's observations regarding the importance of knowledge of paternity" (Pope 1994: 90).

With regard to incest, Arnhart writes that "Aquinas also uses his criteria of mating to assess incest . . . and believes that incestuous mating is contrary to natural law" (Arnhart 2001: 6).[12] Aquinas says that sexual intercourse between persons related by blood is something essentially unbecoming and contrary to natural reason (Aquinas 1270, Qu 154). This kind of conclusion sits well with genetic research that says that incest will create high rates of infant mortality and mental and physical defects.

Another interesting example from Aquinas concerns virginity. A natural inclination toward sex notwithstanding, Aquinas argues that virginity is compatible with the natural law. In his argument Aquinas uses the analogy of eating. Everyone must eat, he reasons, for it is required for each person to stay alive. But in order for *the human race* to stay alive, each person does not need to procreate. The human race is still sufficiently provided for, he thinks, if some people for spiritual reasons choose to abstain from procreation (Aquinas 1270: Qu 152). So while sex is necessary for procreation, particular individuals who wish to follow the natural law need not necessarily procreate.

Overall, Arnhart judges that "much (if not all) of what Aquinas said about the natural inclinations supporting natural law would be confirmed by modern biological research" (Arnhart 2001: 28). Because Aquinas distinguishes natural from divine law by saying that the natural law is known through natural reason while the divine law is not, Arnhart says that "today the religious believer and the Darwinian scientist, differing as they do in their worldviews, can each look to the laws of nature as a ground of common human experience that can be known by natural reason alone" (Arnhart 2001: 32). This is similar to how Ruse argues that a Darwinian can be a Christian. But it also explains how when Wilson refuses to accept the distinction between natural law and divine law by regarding both of them as simply theological, Wilson would conclude that they are both erroneous in comparison with biological science.

7.3 MODELS OF EVOLUTION AND NATURAL LAW ETHICS

7.3a Darwin and Natural Law Ethics

The background theism that is the lynchpin for Aquinas' natural law ethics and the ultimate source of its normativity is obviously absent in Darwin's account of ethics, because Darwin works with a strictly naturalistic perspective. Even so, Arnhart says Darwin's account is compatible with a supernaturalistic "biblical religion" approach (Arnhart 2004a: 204). Arnhart sees Darwin's approach to ethics as situated in the broad natural law tradition that grew from biblical roots. "Darwin's idea of a moral sense rooted in human nature," says Arnhart, "belongs to a tradition of moral naturalism that includes the idea of natural law as elaborated by Thomas Aquinas and other Scholastics. And that idea of natural law is the moral expression of the biblical doctrine of creation" (Arnhart 2004a: 204).

Another contrast between Aquinas and Darwin is that Aquinas appears to hold that ethics is rooted in an invariant human nature. Darwin, though, surely cannot view human nature as invariant, since natural selection guarantees that species evolve. On the other hand, though, Darwin does write as if human nature as we know it today is stable. In speculating about future generations, he predicts that social instincts would grow even stronger and there wouldn't be as much of a struggle between the higher and lower impulses, as he calls them (Darwin 1871: 104).

The social instincts are so deeply rooted in human nature that Darwin cannot foresee human beings losing that dimension. This seems very similar to Aristotle's and Aquinas' view that human beings are social animals by their nature. Aquinas even says that "fairness in buying and selling and the like without which men could not live in society . . . is a part of the law of nature because man is by nature a social animal" (Aquinas 1270: Qu 95).

The golden rule, an ethical principle that natural law ethics incorporates, Darwin sees as consistent with his evolutionary account of human nature (Darwin 1871: 75, 106). As we have seen, Darwin says he has tried to show that social instincts, intellectual powers, and habits lead to the golden rule, which he regards as lying at the foundation of morality (Darwin 1871: 106).

Let's consider Darwin and the key elements of natural law ethics. Most of the four natural inclinations, of self-preservation, procreation, sociability, and knowledge, are consistent with Darwin's observations about human nature. The main ethical principle of natural law ethics says to preserve the fundamental human goods we perceive through reflection on our natural inclinations: we ought to preserve ourselves, procreate, be sociable, and seek knowledge. For the natural law tradition, this main ethical principle is simply hard-wired into human nature.[13] Darwin does not doubt that there is a human inclination toward procreation; he speaks of the "senseless practice of celibacy" (Darwin 1871: 96). He speaks of an "instinct of self-preservation," as well as "social instincts" (Darwin 1871: 71, 87, 90). With regard to a natural inclination toward knowledge, Darwin does describe love of truth as a virtue, and he says humans naturally crave to understand; yet he is puzzled why truth is more highly appreciated by some tribes than by others (1871: 65, 94, 100).

Just as Aquinas discusses situations in which fundamental goods can come into conflict, Darwin, too, refers to situations in which our inclination toward self-preservation and our social inclination to help others come into conflict (1871: 87). At times, Darwin says, we are able to "disregard the instinct of self-preservation" (Darwin 1871: 87).

Apart from the details of the individual natural inclinations, what is most significant here according to Arnhart is that the natural inclinations that Aquinas and the natural law tradition had identified are supported by Darwin's biological account.

> Darwin explains how the natural inclinations that lead to the moral sentiments could have been implanted in human nature by natural selection in the course of our evolutionary history. In doing this, Darwin strengthens the tradition of natural law reasoning by deepening its biological foundation. (Arnhart 2001: 15)

In broad outline, the natural law ethical directives are perfectly in accord with Darwin's account of ethics. And so by Darwin providing a "biological explanation for how the moral sense could be rooted in human nature," Arnhart explains that Darwin has provided logical space for the religious natural law tradition that says that "there is a natural sense of right and wrong that is implanted in the human heart by the Creator" (Arnhart 2004a: 209).

7.3b Dawkins and Natural Law Ethics

Craig Boyd (2004) discusses Dawkins and natural law ethics and argues that natural law ethics can accommodate Dawkins' model of evolution. We saw that Arnhart claims that Aquinas' account assumes that there is a continuity between humans and other animals. Additionally, humans have characteristics that go beyond their animal nature. Boyd follows up on this aspect of Aquinas' approach and compares Dawkins with Aquinas.

> In Dawkins' work a structure—that is to say, the genotype—functions in a profoundly similar way to the manner Aquinas sees the vegetative powers of the soul. It is a principle of organic life that humans share with all other forms of life. Dawkins contends that the only "telos" operating in the genotype is self-replication. One may say that this certainly plays a part in all biological organisms, but it is not the only possible teleology, at least from a philosophical perspective. (Boyd 2004: 228)

Boyd's point is that while Dawkins' and other biological approaches can "provide important elements for a natural law morality," those biological explanations cannot explain all human behavior, "especially those that contravene basic biological impulses" (Boyd 2004: 233). Some areas of "natural law morality," says Boyd, transcend biological explanations (2004: 235). Technically, though, as Arnhart (2008) has pointed out, Boyd is here confusing natural law with divine law. For Aquinas, no area of natural law of morality would transcend a biological explanation, for natural law morality is known by human reason, just as biological explanations are. Divine law morality, on the other hand, *does* transcend human reason. We saw above that Aquinas regards divine law as necessary over and above natural law, because only divine law can direct human beings to *eternal* happiness.

Other similarities between ethics understood in Dawkins' evolutionary perspective and the natural law perspective are that humans have a shared moral sense, there are some moral universals, and that moral progress is possible (§4.3b). Dawkins maintains that with ethics "there is a consensus about what we do as a matter of fact consider right and wrong: a consensus that prevails surprisingly widely" (Dawkins 2006a: 262). This is because humans have a shared moral sense, Dawkins claims. The natural law tradition agrees that there is a shared moral sense, moral universals, and that moral progress is possible. The shared moral sense is what Paul refers to when he says "The natural law is written on the human heart" (*Romans* 2: 14–15). And precisely because of a shared moral sense refined over time, this makes a plausible story for moral progress.

Also, Dawkins, just as Darwin before him, acknowledges that human nature possesses many different natural inclinations. Bearing in mind Aquinas' identification of at least four natural inclinations (self-preserving, propagation, sociability, and knowledge), in the following passage notice how Dawkins unwittingly acknowledges three out of four of these inclinations, and claims that they were created by natural selection.

> Natural selection, in ancestral times when we lived in small and stable bands like baboons, programmed into our brains altruistic urges, alongside sexual urges, hunger urges, xenophobic urges and so on. (Dawkins 2006a: 221)

What he calls "altruistic urges" points to sociability; "sexual urges" points to propagation; and "hunger urges" points to self-preservation.

But what is the ultimate explanation for why human beings possess these urges? Dawkins, as a metaphysical naturalist, will entertain nothing deeper than natural selection. About sexual urges, Dawkins says:

> Sexual desire . . . is a strong urge which exists independently of its ultimate rationale. (Dawkins 2006a: 221)

Aquinas could not agree that sexual desire exists *independently* of its ultimate rationale. For Aquinas, sexual desire exists, and it exists because it serves a role for God's eternal law and natural law. As a metaphysical naturalist Dawkins rejects eternal law, divine law, as well as natural law, and so from his perspective it makes sense to say that sexual desire exists today in human nature, but it exists independently of its ultimate rationale for which natural selection had created it. For metaphysical naturalists it would seem to be up to human beings themselves to decide what rationale makes sense to them about sexuality. Again, although for theologians sexuality is thought to have a metaphysical significance, for naturalists it does not, since sexuality is only an accidental creation of biological evolution; it can serve the role of reproduction but also pair bonding, as Wilson had pointed out, and perhaps other roles as well (social ordering perhaps).

As we have seen, Dawkins emphasizes how the ethical dimensions of human life that he describes can be understood and explained independently of religion. For Dawkins "We do not need God in order to be good—or evil" (Dawkins 2006a: 226). Explicitly, he says, "All I am trying to establish for the moment is that we do not, as a matter of fact, derive our morals from scripture" (Dawkins 2006a: 243). An interesting thing about natural law ethics is that its proponents agree that we need not necessarily consult the Bible to understand ethics, since the natural law is written on the human heart. So even though Dawkins might think he is ruling out theistic ethics, he really isn't. Natural law ethicists could agree that there is a genetic basis for natural inclinations. As Boyd and Arnhart claim, evolution—even as understood by Dawkins—could support aspects of natural law theory.

7.3c Gould and Natural Law Ethics

Gould would be skeptical of natural law reasoning, just as Wilson is and Dawkins would be. Natural law ethics is a normative ethic. Gould would not accept "natural law" as a normative ethical concept. He does occasionally refer to "nature's laws" and "laws of nature" (1999). But he does not view them as normative terms. For Gould, nature just is, and ethical implications or conclusions cannot be drawn from laws of nature.

Given that nature "is what it is," Gould would also reject the overall approach of the natural law tradition that says that ethics depends on *human* nature and human natural inclinations. For Gould, ethics and ethical principles would have to be created by human beings, not a part of reality waiting to be discovered. Like Wilson, Gould would consider the suggestion that there is a natural law directive to value basic human goods as a misguided and erroneous attempt to derive an ethical principle.

On the other hand, though, even Ruse claims that a natural law "type" ethic is possible from an evolutionary perspective. In this vein, Gould has stated that the golden rule is a valid ethical principle with which he agrees (1993: 50–51). And on the topic of the stability of human nature, we have seen that Aquinas probably sees human nature as more fixed than an evolutionist could agree with, yet Gould, just as Darwin, makes the case that the human species is predictably stable, a point we have seen in previous chapters, which is important for a natural law approach.

> *Homo sapiens* has been stable for tens of thousands of years, and any proper understanding of macroevolution as a speciational process must yield this very expectation. . . . The only sensible biological prediction about human futures envisions continued stability into any time close enough to warrant any meaningful speculation. In any case, cultural change, in its explosive Lamarckian mode, has now so trumped biological evolution, that any directional trend in any allelic frequency can only rank as risibly insignificant in the general scheme of things. (Gould 2002: 914)

The stability of *Homo sapiens* fits with Gould's punctuated equilibrium model of evolution. Gould explains that

> the last 50,000 years or more of human phenotypic stability becomes a theoretical expectation under punctuated equilibrium, and not the anomaly so often envisaged (and attributed to the suppression of natural selection by cultural evolution) both by the lay public and by many professionals as well. (Gould 2002: 79)

An example of a part of human nature that has been put in place and stabilized by evolution—and relevant to ethics—are "our altruistic tendencies," that Gould says "may have arisen by the same Darwinian route via kin selection" (Gould 1977: 266).

Over all, it does not seem that Gould's model of evolution itself would conflict with a natural law ethics, but the naturalistic metaphysics that Gould maintains is the source of the tension. This becomes clearer as we examine Haught's theistic model of evolution and natural law.

7.3d Haught and Natural Law Ethics

Gould will only accept the concept of "laws of nature" in a nonnormative sense. Haught, as we have seen, defends a layered explanations approach and so he can acknowledge that nature itself can indeed operate in accord with nonnormative laws of nature, but Haught can also claim that there are other aspects to reality that fall outside the purely empirical dimension that science fastens onto. With his layered explanation approach, Haught says he can combine scientific and theistic notions. He points to "remorseless regularity of laws of nature such as natural selection," and asserts that "[l]awfulness at one level of nature's hierarchically emergent structure is essential for the emergence of novelty and indeterminacy at another" (Haught 2000: 102).

Specifically with regard to a natural law perspective, Haught would suggest that an evolutionary view of the universe can help us reinterpret eternal, divine, and natural law. This is what Haught means when he says that Darwin has given a great gift to theology (2000: 22). Darwin has prompted theology to broaden its conception of reality and take seriously the deep importance of the historical unfolding and evolving of the cosmos. The traditional notion of a divine plan will need to be put into the context of an unfolding cosmos, argues Haught. Evolutionists who have described the struggle and waste that happens in nature through billions of years have made the idea of a divinely designed universe quite unbelievable, Haught says (Haught 2000: 106).

So rather than understand nature along the lines of a traditional natural law approach, Haught proposes that "creation . . . would be less the consequence of an eternal divine 'plan' than of God's humble and loving 'letting be'" (2000: 112). He thinks it is more accurate to speak of "God's vision, not plan" (Haught 2000: 190).

With regard to the specific normative dimension of natural law ethics, Haught doesn't have a robust account of human nature in terms of all four categories of natural inclinations. Yet he does make some claims about human natural inclinations that are parallel to what natural law theorists claim: the most central is his notion of "[t]he attractive power of 'the good'" (Haught 2003: 11). Aquinas, too, had identified a natural inclination toward the good and away from evil as the foundational principle of natural law ethics. Another way that Haught has characterized the natural inclination to the good is the imperative of responsibility that we intuitively sense. At the foundation of our consciousness, Haught claims, are the persistent and unavoidable imperatives "be attentive," "be intelligent," "be critical," and "be responsible." The imperative to be responsible is just there, we feel its pull. This is again where Haught's layered explanations approach allows his theism to come in:

> [T]he imperative to be responsible is activated by the mind's *anticipation* of a transcendent goodness that encompasses and grounds both the world and our consciousness. (Haught 2006: 155)

Specifically, one of the natural inclinations Aquinas mentions is the natural inclination to knowledge, and Haught's account incorporates something similar. Haught says there is a desire to know that since ancient times has been identified in each person's intellect, and it is the root of science (Haught 2006: 33).

Haught doesn't understand evil as the intentional destruction of natural goods, but offers an alternative account of evil that makes use of an evolutionary perspective. About evil he says, "Our evil is our systematic refusal to participate in the ongoing creation of the world" (Haught 2003: 175).

Haught reaches the same conclusion as Boyd who said that biological explanations would fall short in capturing all human behavior (Boyd 2004: 233–34). Haught explicitly says that some people's actions and lives "can be accounted for only on the assumption that they have been grasped by, and have responded to, an unconditional goodness" (Haught 2010: 121). For Haught, a God of love acts through "divine allurement" (Haught 2000: 53). The notion of divine allurement seems to straddle the line between natural law and divine law, for it is a *natural inclination* toward a divine or spiritual future. But a divine or spiritual future would not seem to be something that natural reason alone would be able to perceive. Haught can still make use of Aquinas' notion of divine providence, and that there is a natural inclination to the good written on the human heart; Haught will want to emphasize that it is natural to evolve, for evolution itself is part of God's vision for reality.

7.4 CONCLUSION

Natural law ethics has had a long history and has been examined by biologists, philosophers, and theologians. Although the theory has had its detractors, it still has contemporary proponents. Some (but not all) contemporary proponents of natural law ethics argue that evolutionary theory can support various aspects of natural law ethics. In looking at natural law ethics through four models of evolution it does seem that a natural law type ethic can fit into an evolutionary framework and evolutionary theory can support natural law ethics. Making such a case is easier from a theistic perspective, yet it is still possible even from the perspective of metaphysical naturalism.

We have looked at several examples of how an evolutionary view is compatible with a natural law ethic. Arnhart and Boyd make the strongest cases by showing that conclusions reached by biological science today are similar to conclusions that Aquinas reached by using his natural law approach. And we also saw this when we looked at how Darwin's model of evolution bears on natural law ethics, and when we look at Gould's model and what it implies about the stability of human nature and the applicability of the golden rule.

Stephen Pope supplements this case by identifying several "functional equivalencies" among the natural law tradition and contemporary biology. For Pope, the basic idea is that both natural law and biology perceive a natural ordering in human nature. The notion of natural ordering comes close to the priority that people put on the natural over the unnatural. And from this, it is tempting to say, as did Cicero, that right is based on nature, not on opinion. When used carefully, the natural/unnatural distinction may still be useful in providing a partial glimpse into ethics, especially when we understand the natural as having evolved, as Haught emphasizes.

Wilson bluntly states that natural law ethics is erroneous. Yet Wilson, like Dawkins, incorporates elements that are similar to elements of natural law ethics, most obviously the moral sense that is part of human nature, the moral universals that are perceived *with* that moral sense, and the reality of moral progress.

In the next chapter we will explore how the early modern theory of social contract ethics fits within an evolutionary framework.

NOTES

1. See "Human Cloning and Human Dignity: An Ethical Inquiry," by The President's Council on Bioethics (2002), and Pence (1998: 115). Hefner (1997) argues that cloning is perfectly natural.

2. Kolata (2015) describes a recent (failed) attempt by scientists to perform gene editing on human embryos.

3. The U.S. Food and Drug Administration recently approved genetically altered salmon for consumption; the first genetically engineered animal that vendors will be allowed to bring to market (Pollack 2015).

4. On some interpretations of Aquinas' ethical approach, his theism serves as the ultimate source of the normativity of natural law ethics.

5. All quotations from Aquinas are taken from questions he discusses in the First Part of the Second Part (I–II) of his *Summa Theologiae*, unless otherwise indicated.

6. The notion that the good life is a life that takes us closer to God has an obviously important place in Christianity, and Saint Augustine (354–430) is well known for making it a centerpiece of his ethical approach. Augustine writes: "Following after God is the desire of happiness; to reach God is happiness itself. We follow after God by loving Him" (388, chapter 11, verse 18).

7. Augustine, too, acknowledges this biblical notion that there is a law written on the human heart. In a famous passage from his *Confessions*, Augustine writes: "Beyond question, theft is punished by your law, O Lord, and by the law written in human hearts, which not even sin itself can erase; for does any thief tolerate being robbed by another thief, even if he is rich and the other is driven by want?" (397: bk 2, chapter 4).

8. Aquinas has a fairly well-worked out ethical theory; consequently he deals with situations where it is not clear what the natural law requires, so-called moral dilemmas. A natural law ethical approach can refer to the Pauline Principle (it is not morally permissible to do evil so that good may follow, which is also formulated as, the end does not justify the means). As an example of the use of the Pauline Principle, Aquinas considers a situation where one is being forced to consent to committing a sin. So the question may arise, may I kill myself "for fear of consenting to sin"? Aquinas argues no, by invoking the Pauline Principle "evil may not be done that good may come of it" (1270: II–II Qu 64). He concludes that "no one is allowed to kill himself for fear of consenting to sin," for that would be killing one self (committing an evil), hoping that good would come (one would not consent to sin). Another ethical

principle that natural law theorists can use in dealing with particular moral dilemmas is the Principle of Double Effect, which is derived from a passage in Aquinas where he discusses special situations in which "an act can have two effects, only one of which is intended" (II–II Qu 64).

9. Contemporary discussions and defenses of natural law ethics are found in George (1992) and Paul, Miller, and Paul (2001). Two recent defenses of natural law ethics are Alfonso Gomez-Lobo's *Morality and the Human Goods: An Introduction to Natural Law Ethics* (2001), and Craig A. Boyd's *Shared Morality: A Narrative Defense of Natural Law Ethics* (2007).

10. In an article about natural law ethics and evolutionary ethics (1999) the philosopher of science Phillip Sloan situates natural law ethics historically. Sloan writes: "An ethics founded on the assumption of a universal 'human nature,' incorporating a strong biological component, was not a novelty of the Enlightenment. It had been a primary premise of the preceding natural-law tradition" (Sloan 1999: 54).

11. Since natural law is known through human reason, Boyd puts it this way: "The natural law prohibition on murder can therefore be understood as reason's ability to see the necessary relationship between the principle of nonmaleficence and social cohesion. Normative judgments follow from the synthesis of biological impulses and reason's insight into the complexities of human relationships and the various contingencies affecting social order" (Boyd 2004: 231).

12. Arnhart develops this case in (Arnhart 2004b).

13. As St. Paul has put it, "The natural law is written on the human heart" (*Letter to the Romans* 2:14–15).

Chapter 8

Evolution and Social Contract Ethics

Although egoism is what we would normally think of as the *opposite* of ethical behavior, *ethical egoism* is a longstanding normative ethical theory that comes in several different formulations. Its central normative ethical principle is that one ought always do whatever is in one's best interest. Social contract ethics is closely aligned with ethical egoism. If ethics are the rules established by a social contract, then a pertinent question is, why should I be ethical, that is, why should I follow the rules of the social contract? Some defenders of social contract ethics (beginning with Hobbes in the seventeenth century) say that the answer is quite simple: following the rules of the social contract is an avenue for satisfying one's interest.

In this chapter I first lay out the main points of social contract ethics, as developed primarily by Thomas Hobbes (1588–1679). I then discuss recent attempts by Ruse, Skyrms, Axelrod, and Kahane to link social contract thinking with evolution. I then examine what four models of evolution imply about social contract ethics.

8.1 HOBBES' SOCIAL CONTRACT ETHICS

Hobbes' social contract ethics relies on a particular view of human nature. For Hobbes, viewing humans as naturally social creatures that have a natural inclination toward knowledge and truth (as natural law theorists do) is naïve and self-deceiving. Hobbes subscribes to the view of human nature that characterizes all human behavior as motivated by self-interest, the position contemporary thinkers call *psychological egoism* (1651: 105). Hobbes also observes that human beings are rational creatures, clever problem-solvers who can figure things out to satisfy their desires. In Hobbes' view, human beings in "the state of nature" (as opposed to "the state of society") have nothing to stop them from following their main inclination to preserve themselves. Since there are no ethical laws in the state of nature (ethical laws only exist in societies, argues Hobbes), human beings are totally free to do whatever they require in order to help and protect themselves (Hobbes 1651: 103). And given psychological egoism, every human being has the same tendency. We are all in the same boat; we are more or less equal in this regard.

In social contract ethics, a principle of self-interest (that says one ought always to perform those actions that benefit oneself) is the centerpiece of ethics. This principle is called *ethical egoism*, because it emphasizes how people *should* behave: they *should* do whatever they need to do to help and protect themselves, whether in or out of the state of nature.

But social contract ethics says that in the state of nature people do not really get what they want—a straightforward ethical egoism doesn't work. In a state of nature there will be constant conflict caused by unlimited freedom, and people will not get the security and stability they desire; the condition is a state of war in which everyone is worse off. As clever creatures, human beings devise a way to escape from the state of nature: an enlightened egoism, or "rule-egoism," that is, a social contract, which includes giving up a degree of freedom. Managing the terms of the social contract will, of course, be a balancing act. Eternal vigilance is necessary, for, on the assumption of psychological egoism, we are all selfish, and watchdogs are needed who insure that people are following the rules and not taking advantage of their positions of power in society. Hobbes envisions a strategy of cooperation—agreeing to mutually beneficial commonsense rules of morality (like don't lie, don't steal, don't kill)—as a better scheme for satisfying one's desires for peace, stability, and security. The clearest evidence that social contract ethics endorses commonsense morality is that, according to Hobbes, the golden rule sums up social contract ethics. Hobbes finds no trouble with incorporating a principle traditionally thought to have religious roots into his nonreligious ethic. He will interpret the golden rule with an egoistic slant: why should I not lie to you? Because I don't want *you* to lie to *me*; thus, treat others as *you* wish to be treated (the golden rule). Similarly, he makes use of the concept of "contract" (he often uses the word "covenant" too), which has obvious roots in the Judeo-Christian tradition in which he was immersed, a tradition that believes that there is a covenant between God and his people. The main ideas of social contract ethics include "the state of nature," equality, a social contract, a state of war, law, reciprocity, psychological egoism, liberty/freedom, justice, and ethical egoism. The two key social contract principles are: a principle of self-interest, that is, ethical egoism—one ought always to do whatever is in one's best interest; and, a principle of the social contract—one ought to agree to participate in social contracts.

8.2 SOCIAL CONTRACT ETHICS AND EVOLUTION

After Hobbes, many versions of social contract theory have been developed, such as John Locke's (1689) and Jean-Jacques Rousseau's (1762). The most influential version of social contract theory in the twentieth century is Rawls' (1971). Rawls seeks to defend and develop social contract theory by carrying it "to a higher level of abstraction" (Rawls 1971: 11). Instead of focusing on a contract between citizens and their representatives, or between a sovereign and his and her subjects, Rawls focuses on the principles of justice that individuals would agree to follow, if, hypothetically, they were in a position where they didn't know anything about themselves. In the "original position" behind a "veil of ignorance," which principles would I agree would be fair for the society in which I will be living? (One can notice the background egoism here).

The philosopher Peter Wenz describes what Rawls has in mind:

> Rawls . . . asks us to imagine people who are about to start a new society, like the people on the Mayflower before they landed at Plymouth, or intelligent beings on *Star Trek* about to inhabit a new planet. They discuss the rules of their new society. To make sure that these rules are fair to everyone, Rawls wants us to imagine that the people setting up the rules don't know their own personal identities in the new society. They don't know if they'll be rich or poor, male or female, young or old, tall or short, black or white, extremely intelligent or just above average, and so on. (Wenz 2009: 105)

Rawls will sometimes explicitly refer to biological evolution and natural selection (1971: 431–32, 495, 503–4). When he does so, he is careful to point out that an evolutionary perspective pertains directly to human nature: psychological facts (including moral feelings and a sense of justice) and patterns of behavior. Rawls is clear that he is not attempting to *derive* a social contract theory from these facts of human nature. Rather, Rawls first attempts to establish his social contract theory by making only minimal assumptions. Then, he later inquires from time to time whether such a theory is compatible and consistent with what biological evolution tells us. He finds that natural selection would have favored the kind of creature his theory is about, namely, beings who are capable of having moral feelings, a sense of justice, and patterns of behavior such as social cooperation and reciprocity.

The question of whether these aspects of human are *justified* does not come up, Rawls says (1971: 432). This is because even if an "evolutionary explanation" is correct, it is still an open question whether these aspects of human nature are "to be encouraged and supported" (1971: 432). Here he is distinguishing a theory of human nature from a normative theory.

Another question Rawls asks is "whether the principles of justice are closer to the tendency of evolution than the principle of utility" (Rawls 1971: 503).[1] Even if the principles of justice in his social contract theory *are* closer to the tendency of evolution, those considerations, he says, "are not intended as justifying reasons for the contract view" (Rawls 1971: 504). He explains that

> the main grounds for the principles of justice have already been presented. At this point we are simply checking whether the conception already adopted is a feasible one and not so unstable that some other choice might be better. (Rawls 1971: 504)

Rawls' social contract theory is thought by Michael Ruse to fit very well with an evolutionary framework. We have seen how Ruse defends an error theory, connects it to an evolutionary framework, and has also attempted to cast various normative ethical theories into an evolutionary framework. About Rawls' social contract theory, Ruse says: "The way in which Rawls presents his moral theory sounds almost as if it had been prepared by a Darwinian" (Ruse 1998a: 245). This is because to Ruse, Rawls' theory, "sounds like reciprocal altruism in action" (Ruse 1998a: 245; 2001:198). The concept of "reciprocal altruism" is from the evolutionary biologist Robert Trivers (1971). Rawls (1971: 503) mentions Trivers' notion of reciprocal altruism and says he prefers to simply call it reciprocity.

Like most social contract theorists, Ruse doesn't think that an actual and rationally explicit contract is necessary. Whereas one might say that the contract is rationally implicit, Ruse goes even further, and seems to say that the contract is, in effect, biological, not made consciously. Natural selection brings this about because a contract is a way to maximize an individual's interests (Ruse 1998a: 246). Without this kind of "Darwinian backing," Ruse thinks that Rawls' social contract theory "just leaves us dangling" because it doesn't tell us "why our sense of fairness arises in the first place" (Ruse 1998a: 244, 245). This is a bit misleading because, as I said above, Rawls does entertain the live possibility that our sense of fairness is a product of natural selection.

But Ruse believes that an evolutionary perspective provides a background explanation that is lacking in social contract theory itself. Ruse says that "the strength of the Darwinian position is that it tells us why our sense of fairness arises in the first place," and why we take morality seriously (Ruse 1998a: 245–46).

Brian Skyrms is another philosopher who, in several books including *Evolution of the Social Contract* (1996), has investigated evolution and social contract theory. Although much of the social contract tradition emphasizes rational egoism, according to Skyrms: "There is another tradition—exemplified by David Hume and Jean-Jacques Rousseau—which asks different questions. How can the existing implicit social contract have evolved?" (Skyrms 1996: ix). By working with models of evolutionary dynamics and evolutionary game theory, Skyrms aims to answer this question. Game theory has been traditionally framed in terms of rational decision, but Skyrms believes that game theory can successfully be framed in terms of evolution (1996: 102). He hasn't presented a full theory of the evolution of the social contract, though; but concedes that he has only pointed to some of its elements (Skyrms 1996: 105). Skyrms' contribution to social contract ethics is qualified in the following way. He maintains that his "aims are explanatory rather than normative" and that the concerns of his book are "descriptive rather than prescriptive" (Skyrms 1996: xi, 108). The bulk of Skyrms' work in this area is explanatory rather than normative; he is seeking an answer to the mechanism issue, so his contributions may not seem to be directly relevant to this chapter, which focuses on social contract *normative ethics*. Nevertheless, concern with the interactive dynamics of biological evolution, cultural evolution, and learning provides some interesting *constraints* on the possibilities of how humans might live; he thinks (Skyrms 1996: 109). This is important for ethics, if we understand ethics as the study of possibilities of how one might live (Skyrms 1996: 109). In the work *The Evolution of Cooperation* (1984), the political scientist, Robert Axelrod, has also worked with game theory and evolution. Axelrod's main question is: "Under what conditions will cooperation emerge in a world of egoists without central authority?" (Axelrod 1984: 3). Axelrod says that Hobbes pessimistically answered this question. In Hobbes' view, Axelrod says, "Before governments existed, the state of nature was dominated by the problem of selfish individuals who competed on such ruthless terms that life was 'solitary, poor, nasty, brutish, and short'" (Axelrod 1984: 4). Axelrod argues against Hobbes' view that central authority is necessary in order to enforce the social contract. No centralized control is necessary, even between antagonists (Axelrod 1984: 69). Axelrod aims to show this by employing variations of the prisoner's dilemma game. Axelrod explains:

> In the Prisoner's Dilemma game, there are two players. Each has two choices, namely cooperate or defect. Each must make the choice without knowing what the other will do. No matter what the other does, defection yields a higher payoff than cooperation. The dilemma is that if both defect, both do worse than if both had cooperated. This simple game will provide the basis for the entire analogy used in this book. (Axelrod 1984: 7–8)

In terms of social contract theory, the game illustrates that, as a rational egoist, it may make sense to "defect" from the social contract; but since the other prisoner—also a rational egoist—will reach the same conclusion, the net result will be that both defect and both will do worse than if both had cooperated. Essentially, when everyone rationally decides to defect, that puts everyone into the state of nature. On the other hand, by cooperating (agreeing to a social contract), each individual would be better off. Once the players realize this, they will not need a central authority to tell them to agree to the contract, they will abide by the contract because it is in their enlightened best interest. But Axelrod, Skyrms, and Ruse even believe that *non*rational creatures can evolve cooperative schemes. Axelrod, for instance, says that cooperation can get started between nonrational beings, since "the evolutionary process allows the successful strategies to thrive, even if the players do not know why or how" (Axelrod 1984: 173). Skyrms also advises that we should remember that even bacteria have developed and maintain effective signaling systems (2004: 55).

Axelrod adds that even though foresight is not necessary for cooperation, it can be helpful; and he provides advice to someone in a prisoner's dilemma (1984: 109). "Don't be envious. Don't be the first to defect. Reciprocate both cooperation and defection. Don't be too clever" (Axelrod 1984: 110). As these are recommendations for what we should do—if we seek to do well in an arena of competitive egoists—it appears that Axelrod ventures into normative contractual thought. Axelrod moves beyond models of evolutionary dynamics that aim to provide explanations of *how* the existing implicit social contract *has evolved*, as Skyrms offers, to normative suggestions about which cooperative strategies are successful in the creation and maintenance of a stable social contract.

In his book *Contract Ethics* (1995), the philosopher Howard Kahane also brings together social contract ethics and evolutionary thought. Kahane says:

> Agreement is the moral binder . . . the theory to be proposed is essentially a contractual one, based on the Hobbesian idea that our (admittedly imperfect) ability to make and keep cooperative agreements is the chief reason why, over the years, life has become less nasty, brutish, and short. (Kahane 1995: x)

Kahane views the shortcoming of Hobbes' contract theory in Hobbes' "insufficient attention" to the "ordinary human desire to keep fair contractual agreements" (Kahane 1995: x). This desire, Kahane argues, is put in place by evolution. "Evolutionary forces," he says, "have generated not only specific moral sentiments such as those favoring the keeping of fair bargains, but also a more general disposition inclining us to accept and conform to in-group customs and standards" (Kahane 1995: 15). I don't see why Rawls could not also accept that these moral feelings have their source in biological evolution. Kahane observes that for some, the relationship between a biological theory and an ethical theory is vexing. He is careful to point out that he is not claiming that evolutionary theories can "*prove* anything concerning the bindingness of moral obligation but rather that they can, and do, provide very good reasons for believing that most of us should be well stocked with sentiments tending to increase profitable cooperative ventures, sentiments very much of the kind that have generally been thought of as moral" (Kahane 1995: 16). This is the same kind of qualification we saw that Rawls makes, that between a theory of human nature and a theory of moral justification. For Kahane, an evolutionary perspective coupled with a social contract ethical theory together yield a satisfying account of normative ethics. Like Rawls, Kahane thinks that as a normative ethical theory, social contract ethics doesn't make any extravagant claims about reality; the theory only makes modest assumptions. In social contract ethics, Kahane says there are no "natural or objective rights or liberties of any kind. On the contractual theory being espoused here, rights are generated in the same way as are obligations—by fair agreements" (Kahane 1995: 49). But in addition to social contract ethics being a modest, yet plausible and powerful ethical theory, it also fits with a scientific view of human nature. Kahane's ethical theory appeals to "scientific theories, in particular . . . the theories evolutionary biologists recently have proposed concerning reciprocal altruism" (Kahane 1995: 141–42). For Kahane, these evolutionary theories imply that "the moral side of human nature evolved because it enabled our distant ancestors to gain the tremendous advantages of cooperative, reciprocal behavior" (Kahane 1995: 142). Further, "because the benefits of cooperative behavior are so great, we must cooperate with each other even though we are each other's most serious genetic competitors" (Kahane 1995: 142). Social contract ethics can take account of "the various tugs on human behavior in terms of the categories evolutionary biologists have proposed in recent years" (Kahane 1995: 142). And this is significant for Kahane, because he believes we must "make our moral theories take account of this way of looking at the human condition" (Kahane 1995: 142). Axelrod believes that his research with computer simulation tournaments that were designed to identify the most effective strategies for winning in iterated prisoners' dilemma situations has furnished him with ways to promote cooperation among human beings (1984: chapter 7). Kahane, although confident in *his* understanding of contract ethics, would perhaps look upon Axelrod's suggestions as laudable. Writes Kahane:

> Whether we learn how better to fairly cooperate with each other or instead use our ever increasing power in unfair competitive activities that produce greater and greater destruction and misery . . . constitute the most important and most interesting question concerning the future of the human species. (Kahane 1995: 143)

I now turn to evolution as it has been characterized by Darwin, Dawkins, Gould, and Haught, in order to get a broader picture of social contract ethics from an evolutionary perspective.

8.3 MODELS OF EVOLUTION AND SOCIAL CONTRACT ETHICS

8.3a Darwin and Social Contract Ethics

In *The Descent of Man* Darwin shows his interest in the origin of ethical duties, and he thinks he can grasp their origin from the perspective of natural history (1871: 70–71). Hobbes' answer to the question about the origin of duty is the social contract. Darwin, though, mounts the case that natural history has provided humans with social instincts and feelings of sympathy, and these are the important building blocks for our ethical duties. Darwin's emphasis on the social nature of human beings gives a different portrait of human nature than what Hobbes provides. While Darwin is familiar with moral philosophers who say that "the foundation of morality lay in a form of selfishness," Darwin is not convinced by such accounts (Darwin 1871: 97). Like Hobbes, Darwin observes in human beings an instinct of self-preservation (1871: 75). Selfishness ultimately adds to the power of sympathy, Darwin says, "for we are led by the hope of receiving good in return to perform acts of sympathetic kindness to others" (Darwin 1874: 50). Darwin gives a description of a "bad man," and argues that such a person could nevertheless accept enlightened egoism. This could be due to fear of punishment, he says, or even the realization that his long-term interest would be served by regarding the good of others (Darwin 1871: 92).

An enlightened egoism would suggest a strategy of cooperation. Darwin notes that higher animals engage in reciprocal behavior. He says, "The most common mutual service in the higher animals is to warn one another of danger" and social animals perform little services for each other, and hunt in packs and mutually defend each other (Darwin 1871: 74–75). Darwin views cooperation as a natural strategy, since animals that live socially would need to defend themselves and attack enemies in groups; they would be committed to the group and obedient to a leader (Darwin 1871: 79).

An agreed-upon set of rules is significant, Darwin says, for "No tribe could hold together if murder, robbery, treachery, etc., were common; consequently such crimes within the limits of the same tribe 'are branded with everlasting infamy'" (Darwin 1871: 93).[2]

But how do animals (or humans) arrive at this strategy of cooperation? Darwin sees it as instinctual, Hobbes as intellectual. Darwin thinks the impulse that leads social animals to associate with each other and aid each other is simply the pleasure or satisfaction one gets from performing an instinctive action (Darwin 1871: 79). In terms of natural history, Darwin says it is probable that the animals were induced to live together along the same lines as how hunger and the pleasure of eating induced animals to eat (Darwin 1871: 80). Darwin views the strategy of cooperation as a function of human nature. As we have seen, because humans are social creatures, thinks Darwin, they would inherit the tendency to be committed to friends and obedient to leaders, for these traits are common for social animals (Darwin 1871: 85). Darwin holds that instincts and sympathy are necessary, but not sufficient, however, for establishing an ethic (1871: 75). Darwin agrees with Hobbes that human reason allows for the development of ethical principles such as the golden rule. But when Darwin discusses the human mind, he emphasizes the continuity between animals and humans: "The difference in mind between man and the higher animals, great as it is," he says, "certainly is one of degree and not of kind" (Darwin 1871: 105). Hobbes, though, emphasizes the difference that rationality makes. And we know that Darwin views the social instincts and intellectual powers as undergirding the golden rule and the foundation of morality (Darwin 1871: 106). Hobbes agrees that the golden rule is the heart of morality, but he says the golden rule sums up *social contract* thinking: don't do to others, what you don't want them to do to you (1651: 122). For Hobbes, the contractual strategy is a rational human contrivance devised by individuals in order to get themselves out of the state of nature. In terms of human nature, Hobbes and Darwin agree that humans have an instinct for self-preservation and that humans use cooperative strategies, such as community-wide shared rules. In terms of normative ethics they would also agree that humans should enter into contracts as a strategy for securing well-being. But they *dis*agree about human nature, too: Darwin does not seem to commit himself to psychological egoism, because he observes strong social instincts in human beings. In terms of ethics, Darwin sees it as a natural outgrowth of social animals while Hobbes views it as a product of human rational contrivance.[3]

8.3b Dawkins and Social Contract Ethics

When contemporary evolutionists attempt to explain social and ethical behavior from an evolutionary perspective, they often speak of the self-interest of the individual organism. Dawkins, as a proponent of genic selection, encourages us to analyze self-interested behavior in terms of genes. Dawkins admits, however, that "at times, gene language gets a bit tedious," so he says that "for brevity and vividness we shall lapse into metaphor" (Dawkins 1989: 45). When writing about aggression, for example, Dawkins says "We shall continue to treat the individual as a selfish machine programmed to do whatever is best for its genes as a whole" (Dawkins 1989: 66). Since biologically, individuals are "selfish machines," Dawkins says that "if you wish, as I do, to build a society in which individuals cooperate generously and unselfishly towards a common good, you can expect little help from biological nature" (Dawkins 1989: 3). You must "teach generosity and altruism," he says, "because we are born selfish" (Dawkins 1989: 3).

On the other hand, though, we have seen (§7.3b) how Dawkins claims that

> natural selection, in ancestral times when we lived in small and stable bands like baboons, programmed into our brains altruistic urges. (Dawkins 2006a: 221)

So we are selfish machines with altruistic urges? If Dawkins really believes we have genuine altruistic urges, then he does not hold psychological egoism as Hobbes does, because psychological egoists say that all human behavior is motivated by self-interest. Perhaps Dawkins, like Darwin, holds that humans have both egoistic and altruistic urges.

Because a self-interested drive has a largely genetic component, the kind of selfishness Dawkins has in mind may manifest itself in counterintuitive ways. A mother caring for her children may appear to be a textbook case of ethical altruism, but for Dawkins, a story of selfish genes would encourage us to take a second look. By investing resources in her children, a mother helps her own genes (Dawkins 1989: 127).

Cooperative strategies, in Dawkins' view, should be looked at in terms of expected genetic payoff. Organisms (and their genes) can often do well with strategies of "cooperative restraint and policing" (Dawkins 1995: 121). Whereas kin selection says that helping behavior will evolve among kin (a mother caring for her children, for example), the biological principle of reciprocal altruism says that natural selection favors biological altruism even among non-kin, since the individual organism will benefit from such behavior. For Dawkins, it is not surprising to find strategies of cooperation among organisms. "Although murder and cannibalism do occur in nature, they are not as common as a naïve interpretation of the selfish gene theory might predict," Dawkins says (1989: 67). "Why is it that animals do not go all out to kill rival members of their species at every possible opportunity?" Dawkins asks (Dawkins 1989: 68). "The general answer to this," he says, "is that there are costs as well as benefits resulting from outright pugnacity" (Dawkins 1989: 68). Such costs and benefits can be quantified with mathematical modeling. Dawkins mentions how the biologist J. Maynard Smith "uses the branch of mathematics known as Game Theory" and how Maynard Smith introduces the concept of an evolutionarily stable strategy (Dawkins 1989: 69). An ESS is not a conscious strategy, but rather, a biological strategy that takes shape among genetic competitors over evolutionary time. As we've seen above (§8.2), this is the area that Skyrms and Axelrod work in.[4]

An ESS comes close to contractual thinking, yet there are important differences; most importantly, that "an ESS is stable, not because it is particularly good for the individuals participating in it, but simply because it is immune to treachery from within" (Dawkins 1989: 72).

Dawkins says that humans can create pacts with each other, and the pacts are to the individual's advantage, and they can see that it is in their long-term interests to adhere to the rules of the pact (Dawkins 1989: 73).

E.O. Wilson, another advocate of genic selection, not only views reciprocity and social contract thinking as *possible* for human beings that have evolved through natural selection, but he views these as *central aspects* of human behavior. Wilson writes:

> Reciprocation among distantly related or unrelated individuals is the key to human society. The perfection of the social contract has broken the ancient vertebrate constraints imposed by rigid kin selection. Through the convention of reciprocation, combined with a flexible, endlessly productive language and a genius for verbal classification, human beings fashion long-remembered agreements upon which cultures and civilizations can be built. (Wilson 1978: 156–57)

This cooperative strategy is rooted in our human nature, Wilson thinks. He finds that "there is in us a flawed capacity for a social contract, mammalian in its limitations, combined with a perpetually renewing, optimistic cynicism with which rational people can accomplish a great deal" (Wilson 1978: 164). Like Hobbes, Wilson acknowledges that human contracts are indeed imperfect, yet they are reliable enough to allow for a stable society in which individuals can attain security.

In Hobbes' view of a social contract, the kind of agreement will not necessarily be an egalitarian agreement, as this is not needed for individuals to achieve security. Evolutionists like Wilson also describe how hierarchies of dominance can work very well, both in the human realm and the nonhuman realm of other mammalian societies.

> Behavioral scientists from another planet would notice immediately the semiotic resemblance between animal submissive behavior on the one hand and human obeisance to religious and civil authority on the other. They would point out that the most elaborate rites of obeisance are directed at the gods, the hyperdominant if invisible members of the human group. (Wilson 1998: 259)

Further, Wilson says that

> countless studies of animal species, with instinctive behavior unobscured by cultural elaboration, have shown that membership in dominance orders pays off in survival and lifetime reproductive success. That is true not just for the dominant individuals, but for the subordinates as well. Membership in either class gives animals better protection against enemies and better access to food, shelter, and mates than does solitary existence. (Wilson 1998: 259)

In sum, Dawkins maintains that we have egoistic drives. Taking into consideration that Dawkins has a genic version of these egoistic drives in mind, we see that the contractarian's claim that human beings enter into contracts as a means of preserving their self-interest fits with Dawkins' picture. Human beings enter into these contracts consciously, thinks Hobbes. Although this strategy that Hobbes describes is not exactly what biologists have in mind with their concept of an evolutionarily stable strategy, in terms of ethics, though, the picture of human nature we get from Dawkins' model of evolution meshes quite well with the main elements of social contract ethics.

8.3c Gould and Social Contract Ethics

Gould's is a naturalistic model of evolution, as viewed by a paleontologist. Since Hobbes takes a scientific and naturalistic approach as well, perhaps the approaches would be consistent. In Gould's multilevel model of natural selection he employs the standard biological concept of reciprocal altruism; he describes it as "a genetic incarnation of the age-old adage: you scratch my back and I'll scratch yours (even if we're not related)" (Gould 1977: 255). In ethics, this is often employed as the basic motto of social contract ethics, with its underlying principle of ethical egoism: "I should scratch your back if you will scratch mine." Gould, after rejecting categorical imperatives, proposes that other moral principles, which are premised upon "desire, negotiation, and reciprocity," have appeared in many cultures in many different eras, and have shown their wisdom (Gould 1993: 50). He is referring to the golden rule, but when he characterizes the golden rule as premised upon desire, negotiation, and reciprocity, he seems to be describing a Hobbesian version of the golden rule, not the more traditional Christian version that, in being other oriented, is in accord with ethical altruism, not ethical egoism.[5] In earlier chapters I discussed how Gould views the influence of culture on human behavior as stronger than the influence of biological forces. Such an approach works well with a Hobbesian social contract ethics. For, where Darwin emphasizes the importance of social sentiments as the roots of ethics (§8.3a), Hobbes emphasizes human rationality. Hobbes views ethics as more a matter of human contrivance than a natural outgrowth of human nature in the human condition.

8.3d Haught and Social Contract Ethics

One of the most attractive features of Hobbes' social contract ethics is that he provides a way of conceiving ethics that allows us to see how purely self-interested individuals could cooperate and agree to basic ethical and legal demands, and thereby form a society. As we have seen above, contemporary evolutionists also maintain that human beings are creatures that—because of their biological nature—are capable of cooperation. Haught is in agreement with these theorists. He views evolution as an inspiring story about cooperation, just as much as a story about competition, waste, and struggle (Haught 2000: 21).

Kahane and Ruse see a gap in social contract theory that evolutionary considerations can fill: an explanation of why humans have the desire to keep fair contractual agreements in the first place. Haught's notion of a fundamental human imperative to "be responsible" sounds a bit like what Kahane and Ruse have in mind here. Kahane maintains, though, that a social contract ethics approach can sit comfortably with a naturalistic evolutionary perspective without having to postulate the notion of *theistic* "critical intelligence," as Haught calls it (experience, understanding, judgment, decision); for naturalists do not see much of an issue in accounting for human critical intelligence. So, as we've seen in previous chapters, Haught's model of evolution will clash on some points with naturalistic accounts of ethics; and since Hobbes' social contract ethics is naturalistic, it, too, will be at odds with Haught's theistic orientation. Haught is firm in his position that ethics cannot be fully naturalized.

Another area of contrast between Hobbes's social contract approach and Haught's model concerns the classical notion that human beings have a natural longing for goodness. Haught believes that "religious traditions nourish the basic longing for goodness" (Haught 2000: 122). But a human basic longing for goodness only makes sense, he thinks, if humans have been grasped by the transcendent. Hobbes, it is quite clear, deliberately attempts to exorcise such ideas from ethics. Hobbes finds attributing to humans such desires as the longing for goodness to be much too idealistic. Another obvious contrast with Haught's model and Hobbes' ethical theory concerns Haught's aesthetic principle. Haught would view social contract ethics as extremely limited in scope, even when it takes contemporary naturalistic evolution on board. Haught has a *cosmic* evolutionary perspective, and social contract theorists simply do not look deep enough, he'd say. Haught's focus is on creation that is unfinished as it moves into the future, so he thinks we should act in ways that leave open "incalculable outcomes" (Haught 2000: 159). Haught's aesthetic principle is extremely broad, while social contract ethics, by contrast, is often criticized as having a rather narrow focus. Critics often say social contract ethics does not capture all of what ethics is about. Commonly, ethics is thought to call for more than obeying laws, and thought to have a deeper purpose than simply an arrangement for egoists to get what they want in the long term. This point of view poses a challenge to social contract ethics, since as an ethical theory one can imagine that its code of conduct simply amounts to a set of laws established by the "social contract." Other critics of social contract ethics employ the developmental psychologist Lawrence Kohlberg's theory of moral development. Kohlberg developed a theory of moral development that outlines several stages of moral development: preconventional, conventional, and postconventional. At the preconventional level human behavior is driven by rewards and punishments. At the conventional level it is driven by observing the rules and conventions of one's society. The last stage—the postconventional stage—is reserved for moral heroes, such as Martin Luther King, Jr. or St. Francis of Assisi, persons motivated by an extreme sensitivity to and concern for others (Duska & Whelan 1975: 79). Critics of social contract ethics say that, at best, social contract theory only captures a conventional stage of moral development. Haught uses Kohlberg's theory, in fact, to further support his case against a naturalistic ethic. Haught believes that "the postconventional stage befuddles the naturalist's dream of providing ultimate explanation" (Haught 2006: 158, 161). This is because Haught thinks that maybe selfish genes can explain preconventional and conventional morality, but he says that truly heroic individuals have been grasped by something imperishable and absolutely good (Haught 2006: 162). A Hobbesian social contract theorist would likely say that Haught's aesthetic cosmological principle is an example of interpreting natural events with rose-colored glasses. Hobbes and other standard social contract theorists focus on how humans must band together to protect themselves and how ethics is simply a survival strategy. The scope of Haught's cosmological principle is much too broad for a social contract theorist to take seriously. A Hobbesian naturalistic ethic hopes to take all the mystery out of ethics and show ethics for what it is, simply a contrived arrangement for egoistic individuals to achieve their desired ends. In our human actions, can we meaningfully be thought to care about how we will participate in the unfolding of the whole cosmos?

Haught has an answer for this skepticism, of course. With his cosmic evolutionary perspective he talks about the different levels of explanation: the Darwinian and cosmic processes have immediate causes and mechanisms, but also an ultimate explanation that sponsors them, namely, God (Haught 2000: 165). For Haught, a social contract ethics with biological emendations of the kind Ruse and Kahane speak of will still fall short. As Haught asks: "While all the interesting things that evolutionists talk about are indeed going on, is it possible also that something even more fundamental is occurring?" (Haught 2000: 10).

Whereas Haught's model of evolution holds out hope and promise (he calls Darwin's evolutionary perspective a gift to theology), Hobbes' naturalistic account of ethics is known for its pessimism (2000: 22). "It seems entirely plausible," thinks Haught "that the universe of contemporary science is more congenial to promise than pessimism" (Haught 2000: 116). Even if some naturalists attempt to incorporate optimism (Haught calls them "sunny naturalists"), he finds such attempts doomed. In Haught's estimation, the sunny naturalists have not truly faced the implications of naturalism, but have only adopted the ethical frameworks that theists have developed over the centuries (Haught 2006: 194).

8.4 CONCLUSION

Can social contract ethics fit within an evolutionary framework? If human biological evolution is part of our worldview, do traditional social contract ethics concepts and principles continue to make sense? It is often remarked that Hobbes's description of the state of nature seems to be quite "Darwinian," in the informal use of the term. It is for reasons like these that Dennett says Hobbes was a sociobiologist, 200 years before Darwin (Dennett 1995: 453).

When we look at the work of contemporary evolutionists such as Dawkins, Gould, Axelrod, E.O. Wilson, and de Waal, we'll notice that when writing about ethics they employ the concepts of reciprocity and cooperation, the logic of the prisoner's dilemma, and the self-interested behavior of the individual organism, all of which are part and parcel of social contract ethics. The ethologist Frans de Waal, for instance, says that if one is to be an ethical altruist and take care of others, one need first take care of oneself (1996: 212). For de Waal, a drive of self-interest is not incompatible with the possibility of genuine altruism. de Waal, in his book *Good Natured: The Origins of Right and Wrong in Humans and Other Animals*, uses "quid pro quo" as the title of his chapter 4. He says, "Once a quid pro quo mindset has taken hold, the "currency" of exchange becomes secondary. Reciprocity begins to permeate all aspects of social life" (de Waal 1996: 154). The easiest place to see the connection between reciprocity and morality, de Waal says, is in the sharing of food (1996: 136). de Waal explicitly uses the language of "social contract" as well as contractual reasoning (1996: 130, 132, 136, 205–7, 209).

> If group life is based on a social contract, it is drawn up and signed not by individual parties, but by Mother Nature. And she signs only if fitness increases through association with others, that is, if sociable individuals leave more progeny than do solitary individuals. We are seeing how social tendencies came

into existence—via a genetic calculus rather than a social choice. Even in our species, which prides itself on free will, we may find an occasional hermit who has opted for reclusion; yet we never encounter someone who has consciously decided to become social. One cannot decide to become what one already is. (de Waal 1996: 170)

Like E.O. Wilson, de Waal says the *kind* of agreement will not necessarily be an egalitarian agreement, although egalitarian arrangements are possible (1996: 123–32). Because of what de Waal sees in chimpanzee behavior, he agrees with the work of cultural anthropologist Christopher Boehm who "was the first to fully develop the idea that egalitarianism is not simply the absence of social stratification but the product of vigilance against excessive individual ambition" (de Waal 1996: 240).[6] Another interesting area in which evolutionary thought bears on social contract thinking is with the so-called free rider problem, which, for some, constitutes a serious objection to social contract theory. Evolutionists, though, don't believe a free rider option is as real as one might think. Says de Waal: "It hardly pays to fake generosity or obey rules only in public. People are too astute at distinguishing such a façade from acts that come from deep inside" (de Waal 1996: 116). Physiological blushing helps to see who is trustworthy. Our ancestors may have benefited from being able to track the reputation of those who may choose the path of free ridership, and also have benefited from "advertising trustworthiness" by sometimes turning red in the face (de Waal 1996: 116).

Even though the work of evolutionists often incorporates elements of social contract thought—sometimes implicitly and sometimes explicitly—some evolutionists who specifically turn to the work of moral philosophy, social contract thought included, often wrongly interpret social contract ethics. E.O. Wilson, for instance, wrongly interprets Rawls (and the nature of moral philosophy generally) when he says: "While few will disagree that justice as fairness is an ideal state for disembodied spirits, the conception is in no way explanatory or predictive with reference to human beings" (Wilson 1980: 287). First, it is simply an error to think that Rawls' theory of justice is meant to be "explanatory or predictive."[7] Second, Rawls' justice as fairness does not depend on characterizing human beings as embodied spirits. On the contrary, Rawls' scheme for justifying his theory of justice works with a modest characterization of human nature, a characterization open to the suggestions of evolutionary biologists. Rawls notes that our concern for justice is grounded in a capacity for "reciprocity" (Rawls 1971: secs. 75–76). The utilitarian by contrast, according to Rawls, "stresses the capacity for sympathy" and relies on unconditional altruistic impulses to make us support the institutions he favors. Such altruistic impulses, Rawls concedes, do exist, but they are far weaker than the dispositions that underlie his own system:

dispositions to reciprocate benefits (Rawls 1971: 501). The point here is that Rawls is interested in defending an account of justice that is fitting for human beings as they really are. Wilson seems to have no appreciation for Rawls' efforts in this regard.

Wilson further maintains that "adding wisdom and insight to the social contract, can come only through a deeper scientific examination of morality" (Wilson 1978: 166). Rawls specifically says that his account of human nature is not *given* by natural science, yet he is open to checking whether his account squares with a scientific account. Nevertheless, Rawls goes a long way in "specifying the terms of social cooperation between citizens regarded as free and equal, and as normal and fully cooperating members of society over a complete life" (Rawls 1993: 18, 20). Given the influence of Rawls' work, it would be strange to say that because he hasn't *relied* on a scientific account of human nature and morality, he hasn't provided insight into the social contract.

de Waal gets Rawls wrong, too; de Waal says, "Rawls goes so far as to present the 'initial situation' of human society as one involving rational but mutually disinterested parties. . . . As if our ancestors did not descend from animals that had lived for millions of years in hierarchically structured communities with strong mutual attachments" (de Waal 1996: 167). It is erroneous to think that Rawls' concepts of the original position and the veil of ignorance depend on a literal origin of such a society, in the same way as thinking that social contract theory depends on a distant foundation day, which is the common objection that Singer has voiced against social contract theory (1981: 3, 23). Actually, this common objection seems to be assuaged with help from evolutionary considerations, since Axelrod has argued that an actual rationally explicit contract is not necessary for the evolution of cooperation; so perhaps that kind of contractual account and the foundation day on which it allegedly depends may not be necessary for a basic ethic to get off the ground. In the light of some of the work of evolutionists, the objection that the social contract is a fictional contract doesn't carry the same weight anymore. An evolutionary perspective does not seem to discredit a social contract understanding of ethics, as Singer alleges.

As for Haught's model of evolution and its implications for social contract ethics, he agrees that strategies of cooperation have a legitimate place in evolutionary history; yet he holds that human beings cooperate in socially beneficial ways because human nature has a longing for goodness; and it is this longing for goodness that for him more plausibly explains the rare individuals who occupy the so-called postconventional stage of moral development. To Haught, social contract ethics, as a naturalistic ethic, is only a partial glimpse into the nature of ethics. In terms of justifying what kind of life we should live, Haught would say that social contract ethics is a bit too mundane, as compared with his aesthetic cosmological principle that says we ought to contribute to the intensification of beauty in the cosmos instead of its deterioration. Overall, an evolutionary perspective provides a fruitful standpoint from which to view the insights and contributions of social contract ethical thought. Although some evolutionists make mistakes when they attempt to make explicit reference to social contract theory, the truer testament to the viability of social contract ethics placed in an evolutionary framework is that many evolutionists make use of contractarian concepts and principles without even realizing that they are the key elements of a well-developed ethical theory: the principle of self-interest, the principle of the golden rule, the principle of the social contract, reciprocity, and cooperative strategy. Yet, contractarian thought may benefit from more acquaintance with evolutionary thought. As I've mentioned, evolutionary considerations may help to deflect common objections to social contract theory, such as the free rider problem, Hobbesian unjustified inegalitarianism, and that a social contract depends on a distant contract day.

Hobbes' social contract ethics is an early modern ethical theory. In the next chapter we will look at deontological ethics, another modern normative ethical theory, one that has proven to be even more popular than social contract ethics, yet has not enjoyed as much popularity with evolutionists.

NOTES

1. The principle of utility will be discussed in chapter 10.
2. In morality, though, Darwin notes that there is plenty of room for superstitions, and absurd and strange rules; see (Darwin 1871: 91, 99, 103).
3. Scanlon's (1998) contemporary version of social contract theory *de*emphasizes the egoistic dimension.
4. Dawkins discusses Axelrod's work in (1989, chapter 12). Mackie (1978: 126–31), in connection with a discussion of Dawkins' model of evolution, mentions the ESS strategy of grudger as reciprocal altruism, and says it thrives, as this model of evolution would predict.
5. de Waal (1996: 136) says, "It is clear why the very first step in the direction of the Golden Rule was made by creatures who began following the reciprocity rule 'Do as the other did, and expect the other to do as you did.'"
6. See Boehm (2000a; 2000b). Also, Marc Bekoff (2004) claims that studies of the evolution of morality can greatly benefit from studying animals beyond nonhuman primates. Play, a behavior observed in many species, implies cooperation and fairness between animals.
7. Singer (1981: 78) points out Wilson's error. Ruse agrees that Wilson's sharp criticisms of Rawls, "are based on total misreadings" (Ruse 1998a: 245).

Chapter 9

Evolution and Deontological Ethics

> That which mingles these pure principles with the empirical does not deserve the name of philosophy . . . much less does it deserve that of moral philosophy (Kant's *Groundwork*, trans. Thomas Kingsmill Abbott, 1895).

In his book *Good Natured: The Origins of Right and Wrong in Humans and Other Animals*, the primatologist Frans de Waal asks, "What is different about the way we *act* that makes us, and not any other species, moral beings?" (de Waal 1996: 111).

The philosopher Christine Korsgaard answers de Waal's question by saying that what makes the difference is that humans can assess, and then adopt, their intentions. Humans have "the capacity for normative self-government, or, as Kant called it, 'autonomy'" (Korsgaard 2006: 112). "It is at this level that morality emerges," Korsgaard says. Korsgaard says she derives her answer from the Kantian tradition in ethics. In this chapter we will consider the normative ethical theory famously developed by Immanuel Kant (1724–1804). Kant asserts that he has identified the one fundamental underlying principle of ethics, which he thinks human beings are aware of in their common sense. The word *duty*, Kant says, carries the essential meaning of what ethics really is, but we need to analyze it more fully, Kant asserts. Once we do so, we can see how for the Kantian deontological tradition, the concept of *autonomy* that Korsgaard mentions turns out to be the most significant feature of morality, which lies deeper than the commonsense notion of *duty*, but is still connected to it. Autonomy basically means *freedom*. And Kant thinks that a shorthand way to describe the subject of ethics is to call it the study of "the laws of freedom" (Kant 1785: 1). The key elements of Kantian deontological ethics include duty (obligation), autonomy, law, dignity, the categorical imperative, person, rights, and universalizability. Deontological ethics is known as a nonconsequentialist normative ethical theory, as we'll see, since it claims that the consequences of an action are *not* what give an action its moral worth.

After I lay out the main concepts of Kant's deontological ethics, I will look at several authors who consider deontological ethics from an evolutionary perspective (Nagel, Korsgaard, Ruse, and Richards). And lastly, I examine deontological ethics from the perspective of four models of evolution.

9.1 THE DEONTOLOGICAL ETHICS OF KANT

Let's start where Kant, in his *Grounding for the Metaphysics of Morals* (1785) says *he* starts—with commonsense notions about ethics. Kant takes it as a given that we are all familiar with "the common idea of duty" (Kant 1785: 2). This is part of the ordinary rational knowledge of morality, he thinks (Kant 1785: 7). Another concept that Kant believes "already dwells in the natural sound understanding" is the notion of a "good will," a will that is good in itself, not good because of what it brings to oneself or brings to others, but simply good intrinsically. The notions of duty and a good will are closely connected, says Kant. He asserts that "the concept of *duty* . . . includes that of a good will" (Kant 1785: 9).

Kant also takes it as a given that people are acquainted with "the common idea . . . of moral laws" (Kant 1785: 2). Just as there are physical laws of nature, there are also laws of freedom, that is, moral laws. The physical laws of nature are the laws according to which everything happens in nature. Moral laws, by contrast, "are those according to which everything ought to happen" (Kant 1785: 1).

With regard to the concept of a law, Kant also takes it as a given that laws are the kinds of things that apply universally. Physical laws are the same everywhere, he thinks. And because they are the same everywhere, then it makes sense to say that they apply absolutely and necessarily. An ironclad physical law of nature will not affect things only somewhat or possibly, but absolutely and necessarily.

Kant carries these assumptions about laws into the area of ethics and moral laws. And again, he thinks the moral law as understood in this way is a commonsense idea. He says that for a law to be morally valid it must be absolutely necessary and apply to everyone (Kant 1785: 2).

Many of these different concepts we have looked at thus far fit together. First, bearing in mind the concept of duty, and what it means to act out of duty and have a good will, then the *non*consequentialist aspect of morality comes out, and we can see how morality has a law-like and absolute character to it. Kant says, "A good will is not good because of what it effects or accomplishes, nor because of its fitness to attain some proposed end: it is good only through its willing, i.e., it is good in itself" (Kant 1785: 7). What makes something morally good, then, is due to the good will itself, not the consequences that the will can bring about.

Kant says that the moral worth of an action does not depend on the realization of the object of the action, but merely on the principle of volition (Kant 1785: 12–13). This today is what is called a nonconsequentialist way of understanding ethics. Kant also points to the contrast between performing an action out of duty and performing an action out of desire or inclination. For Kant "Neither fear nor inclination, but solely respect for the law, is the incentive which can give an action moral worth" (Kant 1785: 44).

If it is the case, then, that the moral worth of an action is not determined by the consequences it brings, but only due to its being done out of duty, with a good will, and out of respect for the moral law, then how can we operationalize this as a normative ethics that gives advice about what we ought to do? Since laws have a universal dimension to them, and at the same time the moral worth of an action comes from the inner willing ("inner principles" he sometimes calls it), thus, Kant reasons that there is nothing else to serve as principle except the universal conformity of the will's actions to law. In short, I should only act when I can also will that my maxim would become a universal law (Kant 1785: 14).

Kant calls this the moral command that necessarily must be followed, the "one categorical imperative" (Kant 1785: 30). From his preface to the *Grounding for the Metaphysics of Morals* we know that Kant is attempting to elucidate the true nature of ethics and spell out the one main principle of ethics. There, he wrote that he was seeking out the supreme principle of morality (Kant 1785: 5).

After Kant applies this principle to several scenarios to show how it can be used to guide us to the morally right thing to do in a situation, he concludes that all duties depend on the one principle, not on the objects of their actions (Kant 1785: 32). In trying to figure out if something is the morally right thing to do, the categorical imperative basically says to examine whether the action can be consistently universalized by one's will, that is, whether the action has universalizability.

The categorical imperative is a pure, law-like moral principle that can point out what we ought to do, regardless of our desires or inclinations or whether the action will make us or others happy; it clarifies that which we have a *duty* to perform.

There are several more concepts that fit into this Kantian deontological theory. While physical laws of nature apply to all beings, the moral law does not. The moral law, as we mentioned above, is the law according to which everything *ought* to happen. The awareness not only that something *is happening*, but that something *ought* to happen is possible only for rational beings. Only rational beings possess the ability to consider whether an action is universalizable or conforms to the moral law. And the ability to decide how one will use one's will is possessed only by rational beings. Only a rational being has the power to act according to its conception of laws; thereby a rational being has a will (Kant 1785: 23).

A rational being is capable of "legislating universal laws" and is "free as regards all laws of nature, and he obeys only those laws which he gives to himself" (Kant 1785: 41). Rational beings, in other words, are free; they have autonomy because they can choose which course of action and which law they will follow.[1]

Another term that Kant uses here is *person*. Rational beings, unlike other kinds of beings, can have a viewpoint; and they can decide what to do. Because of this, they have a moral status, which is conveyed by calling them *persons*. A rational being occupies "the viewpoint that regards himself, as well as every other rational being, as being legislative beings (and hence are they called persons)" (Kant 1785: 43). Again, because rational beings have this unique ability that other kinds of beings do not have, Kant emphasizes their important moral status by using the term *dignity*. Kant marvels at

> the dignity of a rational being who obeys no law except what he at the same time enacts himself. (Kant 1785: 40)

From the concepts of *person* and *autonomy*, Kant formulates another way of stating the categorical imperative.

First he makes the contrast between something being an end in itself and being merely a means.

> [R]ational beings are called persons inasmuch as their nature already marks them out as ends in themselves, i.e., as something which is not to be used merely as a means and hence there is imposed thereby a limit on all arbitrary use of such beings, which are thus objects of respect. Persons are, therefore, not merely subjective ends, whose existence as an effect of our actions has a value for us; but such beings are objective ends, i.e., exist as ends in themselves. (Kant 1785: 36)

With these concepts in hand, Kant formulates another version of the supreme principle of morality (the categorical imperative) that has normative force as well as straightforward applicability:

> The practical imperative will therefore be the following: Act in such a way that you treat humanity, whether in your own person or in the person of another, always at the same time as an end and never simply as a means. (Kant 1785: 36)

Kant asserts that both formulations of the categorical imperative are basically the same (Kant 1785: 42–43). As a way to see how the different concepts of duty, autonomy, dignity, law, as well as the categorical imperative fit together, consider this summary passage from Kant.

> From what has just been said, there can now easily be explained how it happens that, although in the concept of duty we think of subjection to the law, yet at the same time we thereby ascribe a certain dignity and sublimity to the person who fulfills all his duties. . . . We have also shown above how neither fear nor inclination, but solely respect for the law, is the incentive which can give an action moral worth . . . the dignity of humanity consists just in its capacity to legislate universal law, though with the condition of humanity's being at the same time itself subject to this very same legislation. (Kant 1785: 44)

One last ethical concept I will mention here because later deontologists will use it more and more frequently, is the concept of *rights*. Kant uses it less frequently, but it still has a natural place in his deontological ethical theory. To respect a person is to respect a person's rights. That rights have a natural place among the concepts of person and the categorical imperative is clear when Kant writes that "a transgressor of the rights of men intends to make use of the persons of others merely as a means" (Kant 1785: 37).

As we move on to discuss evolutionary perspectives on deontological ethics, it will be helpful to keep in mind Kant's quick shorthand definitions of both morality and duty. Kant's view is that "morality is the relation of actions to the autonomy of the will" and "[d]uty is the necessity of an action done out of respect for the law" (Kant 1785: 13, 44).

To round out this sketch of deontological ethics, we can consider two examples of contemporary Kantian deontologists: Thomas Nagel and Christine Korsgaard. To label them as Kantian deontologists does not mean that they subscribe to *all* aspects of Kant's theory, only that they take on many Kantian elements.[2]

Nagel agrees with much of Kant's characterization of the nature of ethics. For example, Nagel agrees that ethics concerns rational requirements on action. Moral principles state rational conditions and rational requirements on action (Nagel 1970: 3). For Nagel, "The correct morality will always have the preponderance of reasons on its side" (Nagel 1986: 199). And ethics is not about interests, effects, or outcomes of actions (Nagel 1970: 3). Like Kant, Nagel holds that ethics and practical reason are for governing the relations "among actions, desires, and beliefs" (Nagel 1997). This governing is possible only because of freedom; thus, there is a close connection between ethics and freedom (Nagel 1986: 135). Similar to what Kant says, freedom comes from that ability to see ourselves, believes Nagel (1997: 251). Morality itself "is possible only for beings capable of seeing themselves as one individual among others," claims Nagel (1997: 252). Ethics requires that we see ourselves with some detachment (1986: 187). This notion of detachment captures what Kant does when he says he has stripped away all interests, personal and otherwise, and so there is nothing left to serve as a principle for the will except the universal conformity of its actions to law as such. With this line of reasoning, Kant derives the categorical imperative.

Korsgaard, too, thinks that many elements of Kant's theory accurately capture what ethics truly is. Autonomy is the source of duty and obligation, she argues (1996: 93). Autonomy derives from the ability to be self-conscious and reflective. While Nagel uses the term "detachment," Korsgaard calls it "reflective distance" (1996: 113). Korsgaard even incorporates the concept of law.

> The reflective structure of human consciousness requires that you identify yourself with some law or principle which will govern your choices. It requires you to be a law to yourself. And that is the source of normativity. (Korsgaard 1996: 97)

"Being a law to yourself" is what she calls "normative self-government," and that is the distinctive aspect, she argued, in response to de Waal, which accounts for why human beings are moral beings (2006: 112).

9.2 DEONTOLOGICAL ETHICS AND EVOLUTION

In Kant's *Critique of Judgment* he briefly discusses evolution (1790: 309–10). Yet, interestingly (and predictably since Kant was writing in the late 1700s), he is referring to earlier versions of the theory of evolution, not Darwin's evolution by natural selection.[3] Below, we will look at contemporary authors who have considered deontological ethics from an evolutionary perspective.

R.J. Richards says he defends a theory of evolutionary ethics (1986; 1993) (§1.4). He attempts to develop an account that spells out what makes an action moral. Thus, he is offering a normative ethical theory. He argues that an action is morally good not because it has good consequences but because it is intentionally performed by a certain kind of motive (1986: 264–65; 1993: 123). With this, he has included a deontological dimension in his theory of evolutionary ethics; specifically, he says he offers a nonconsequentialist theory of evolutionary ethics.

Richards also contends that evolution has formed a part of human nature according to the criterion of the community good, in that it has operated by kin selection and group selection (Richards 1986: 281). It follows, then, he claims, that a morally good action is one performed with the motive of bringing about the good. I only point to Richards' theory as an example of how deontological elements (in this case nonconsequentialism) can be assimilated into an evolutionary perspective on ethics. Although Richards regards it as a nonconsequentialist feature, the motive is being evaluated by whether it intends to bring about the good of the community, so his theory still seems to have an overall consequentialist structure to it. To see the contrast, we can think of Kant. For Kant, the motive or maxim is not to be evaluated by looking to any outcome for the community, but only in the very act of willing itself—whether the maxim be willed as a universal law or not. Clearly, Kant's ethical theory is a pure version of deontology, unlike Richards' hybrid approach. It remains to be seen whether a purer version of deontological ethics makes sense once we adopt an evolutionary perspective.

According to Michael Ruse, Kantian ethics "can as readily be given a Darwinian backing as can utilitarianism—subject to modifications" (Ruse 1998a: 244). Kant and Korsgaard, one may recall, view morality as built up out of autonomy. Ruse concedes that humans have the kind of freedom that deontologists say we need to get ethics going: "Even though we are clearly part of the causal chain," writes Ruse, "we have some element of choice . . . hence there is a window for morality" (Ruse 1991a: 72). Ruse even says Kant was correct about the categorical aspect of morality, because morality is laid upon us, and we are not free to choose what is right and what is wrong (Ruse 1998a: 259). And, says Ruse, "There is certainly that in Kant's philosophy which attracts the Darwinian" (Ruse 1998a: 262). This is because Kant does not view morality as an external phenomenon, but views it as part of the conditions of rational beings (Ruse 1998a: 262).

Ruse interprets one version of Kant's categorical imperative as the principle that says we should not use others as means (because we don't want to be treated as means) (Ruse 1998a: 244). When Ruse casts the categorical imperative in this way, he believes the categorical imperative squares rather well with reciprocal altruism.

Unfortunately, though, when Ruse adds the phrase "because we don't want to be treated as means,"—the very part that sounds like *reciprocal* altruism—Ruse has departed from the categorical nature of the principle. This, Ruse should realize because he says the principle "is laid upon us, without any 'ifs.'" But Ruse has added an "if" however, that is, if you don't treat me as a means then I won't treat you as a means. Ruse has (accidentally?) turned Kant's deontological principle, one that focuses on doing one's duty for the sheer principle of the thing, into a consequentialist principle, that is, if I don't want to be treated as a means then I ought not to treat you as a means. But this isn't the categorical imperative any more. This is now what Kant calls a hypothetical imperative. For Kant, hypothetical imperatives, which are consistent with consequentialism, are not moral imperatives. Kant rejects consequentialism.[4] With his other version of the categorical imperative, Kant argues that if we cannot will our maxim to become a universal law, then that indicates that the proposed action ought to be morally rejected. This point might be attractive to Darwinians says Ruse because it bears a similarity to what evolutionists term "maladaptive interactions," which Ruse thinks "the Darwinian would think eliminated by selection" (Ruse 1998a: 262). Because of the above points, Ruse comments that "in Kant's moral philosophy there is much in spirit which makes it a plausible pre-evolutionary precursor" (1998a: 262). On the other hand, though, upon more analysis Ruse thinks there are serious tensions between Kantian ethics and an evolutionary approach. This is because Kant "desired a foundation that is alien to Darwinism. In particular . . . he wanted a necessity to moral imperatives foreign to the modern evolutionists" (Ruse 1998a: 263). By contrast, says Ruse, a "Darwinian . . . ties morality tightly to contingent human nature" (Ruse 1998a: 263; 1991a: 69–70). Ruse holds that Darwinians frame a moral sense of right and wrong as simply "a cost-effective way of making humans co-operative," and this moral sense was put in place by evolution (Ruse 1998a: 264). In a way, this is similar to Richards' overall consequentialist framework: a moral sense is evolutionarily important because of the consequences it brings, namely, it aids cooperation, which fosters human survival.

Further, says Ruse, if Kantianism means indifferent universalization, then biology is against it, since we should have stronger feelings and a stronger sense of moral obligation for kin than strangers (1991a: 69). For Ruse, it does not make evolutionary sense to think that the categorical imperative (morality as Kant has described it) be binding on all rational beings. There are many ways to accomplish something (1991a: 69). Ruse finally concludes that the "spirit of Kantianism is antithetical to the spirit of Darwinism" (Ruse 1998a: 265). Philosopher John Teehan (2003) is also skeptical of Kantian ethics when put in the light of evolution. Teehan argues that Kant's understanding of human reason is at odds with an evolutionary understanding of the mind (and emotions). Teehan recommends that "we carry on the work of moral philosophy in a way that takes into account the whole human being, not some idealized 'rational agent' or transcendent self; that we work with a picture of human nature supported by the best findings of science" (Teehan 2003).

Joshua Greene (2008) has actually used the findings of science to make a case against deontological ethics. He has used empirical data and evolutionary considerations to make the case that deontologists *claim* and *think* they are using rational thinking when doing ethical reasoning and making ethical judgments, but empirical data (brain scans, neuroimaging, and various experiments) indicate that the parts of the brain used in making deontological ethical judgments (as opposed to making consequentialist ethical judgments) are the parts of the brain associated with *emotional response* (the amygdala, the posterior cingulate cortex, and the medial prefrontal cortex), as opposed to brain regions associated with *cognition* (the dorsolateral prefontal cortex and inferior parietal lobe) (Greene 2008: 42–44). Moreover, Greene argues that deontological judgments are really emotional "alarm-processes," which are responses triggered by ethical situations, rather than genuine cognitive processes. And this is borne out, he argues, by experiments that show a correlation between deontological moral judgments in response to "up close and personal" moral dilemmas involving harm to others, that differ from consequentialist moral judgments that seem more plausible with distant moral situations involving harm to others. These considerations make sense, he thinks, if we take an evolutionary perspective, because in evolutionary time our species has had more experience with up close and personal violence and harm, than with impersonal and distant harms caused by our actions. And natural selection would favor the use of emotional triggers as a way for humans to handle harmful situations (Greene 2008: 43).

Does this kind of empirically grounded experimental philosophy refute Kantian deontological ethics? Does it at least refute Kant's account of human nature, as Teehan seems to think?

One of Teehan's main points is that Kant's notion of a "rational being" does not sit well with an evolutionary understanding of human nature, because in the course of human evolution our emotions and social instincts predated our intelligence, and the concept of a pure "rational being" does not make evolutionary sense because today our cognitive and emotional processes are intertwined. Greene's research bears this out as well. So there are definitely strands of Kant's view of human nature that sit uncomfortably with contemporary evolutionary biology and neuroscience.

The implications of this empirical research for Kantian *ethics* are more complicated, however. Let's consider Teehan's point with regard to the ethical components of Kantian deontological ethics—duty (obligation), autonomy, law, dignity, the categorical imperative, person, and rights. For creatures that had not evolved a capacity for reflection and a capacity for autonomy, then it is true to say that Kant's ethic would not be useful for those creatures. But once a creature has reached the threshold of rationality and it now possesses the capacities for reflection and autonomy—the capacity to assess and adopt intentions—as Korsgaard puts it, then the ethical components become useful. Having crossed the threshold of rationality, the being can perceive its own autonomy, choose to act from duty, conform to universal law, follow the categorical imperative, perceive its own and others' dignity, and consider persons' rights.

Simply because Kant recommends that we should strive to be idealized rational agents does not mean that Kant believes we *are* ideal rational agents.

Kant is well aware that people experience emotions and are led by emotions. But in terms of a normative approach, the question still remains, what *should* people allow themselves to be led by? The best case for supporting a Kantian ethic is to emphasize it as a normative ethic, not as a description of how ordinary humans make moral judgments.[5] I think this is how the contemporary deontologists Nagel and Korsgaard can defend it.

These points apply to Greene's case about deontological ethics. He is concerned that deontologists *claim* and *think* they are using rational thinking when making ethical judgments, but the underlying story is that their process of ethical reasoning heavily depends on emotional triggers and alarms. Like Teehan, Greene characterizes Kantian ethics as a theory about how ordinary humans make moral judgments. Greene is probably correct that people do not make moral judgments by only using their cognitive abilities. But if we take our question to be about what we should do and how we should try to determine if an action is right or wrong, then a normative ethic that (i) says we should respect persons and their rights, and their autonomy, and their dignity; (ii) says we should think about acting out of duty, and offers the categorical imperative as an ethical principle that helps us to do that because it recommends acting on disinterested principle, not on emotion, then a Kantian deontological ethic still seems serviceable, practical, and ethical.

Research done on psychopaths reveals that they have no trouble carrying out abstract reasoning, yet they do show deficits in experiencing moral emotions. Some have taken this as a refutation of Kantian ethics, since if Kant's theory is correct, then supposedly one need not experience emotions to reach ethical conclusions, but only use one's reason in employing the categorical imperative. But this argument assumes that merely because a person has the capacity to engage in abstract reasoning, then the person will generate defensible ethical conclusions. Why should we assume that? As mentioned above, Kant does recommend that we should strive to be idealized rational agents, but he does not assume that people *are* ideal rational agents. Also, are psychopaths familiar with the formal ethical principle that Kant calls the categorical imperative? And even people who are familiar with the principle can still use the principle poorly, such as the Nazi Adolph Eichmann who famously stated that he was familiar with Kant's principle and used it on a daily basis throughout his life (Arendt 1963).

A different kind of response to Greene's case comes from the philosopher Mark Timmons (2008), who suggests that deontologists can acknowledge Greene's empirical data, and yet bolster their case by joining a deontological normative ethic with an expressivist metaethic. Although traditional deontological normative ethics is linked to a rationalist metaethic—as is Kant's—it is feasible and recommended by Timmons to develop a deontological normative ethic on an expressivist metaethic. An expressivist metaethic coupled with a deontological normative ethic would accommodate Greene's findings about emotions and ethics. Given the data about brains and evolution that Greene has brought to the table, Timmons' suggestion might be a way to bring a deontological ethic more in line with an evolutionary perspective. Timmons' point is that a deontological ethic does not necessarily have to be a Kantian-styled deontology that emphasizes ethical rationalism. Michael Slote (2001: 79–87) also argues that there is promise in a "sentimentalist deontology."

The clash between Kantian ethics and evolution is apparent in Nagel's claims about ethics and biology. In 1978 Nagel published an essay called "Ethics as an Autonomous Theoretical Subject," which he later republished with the title "Ethics Without Biology" (1979b). Nagel's main argument is that ethics has a subject matter just as math, physics, biology, etc., have subject matters. And these subject matters are investigated with rational methods. Therefore we shouldn't look for a biology of ethics any more than we would look for a biology of math (1978: 196).

In a later work, Nagel puts the point this way:

> In my opinion, someone who abandons or qualifies his basic methods of moral reasoning on historical or anthropological grounds alone is nearly as irrational as someone who abandons a mathematical belief on other than mathematical grounds. (Nagel 1997: 244)

As with other subjects of rational investigation, ethical inquiry makes progress, albeit slow progress, claims Nagel (1978: 197–98). "Ethics progresses," he says, "by subjecting impulses to examination" (Nagel 1978: 198).

Those theorists, then, who wish to offer an evolutionary explanation of ethics or to provide biological or evolutionary foundations for ethics, are misguided, he argues. Nagel summarizes his position as follows:

> The creatures who engage in the activity of continually developing methods in response to the problems that arise in the subject of ethics have the capacity to perform the reflective and critical tasks involved . . . this capacity is presumably somehow a function of their organic structure. But it would be as foolish to seek a biological evolutionary explanation of ethics as it would be to seek such an explanation of the development of physics. The development of physics is an intellectual process. (Nagel 1978: 200)

Just as we saw with Korsgaard, Nagel takes a key to ethics to be the intellectual capacity of autonomy, which involves the ability to see ourselves with a reflective distance. This intellectual capacity presumably has roots in evolution, thinks Nagel, but once in place, it should follow rational and logical processes, not physical and biological ones.

> Presumably the human intellectual capacity that has permitted this extremely rapid process to occur was in some way an effect, perhaps only a side-effect, of a process of biological evolution that took a very long time. But the latter can provide no explanation of physical theories that is not trivial. What human beings have discovered in themselves is a capacity to subject their pre-reflective or innate responses to criticism and revision, and to create new forms of understanding. (Nagel 1978: 203–4)

Perhaps an ethic like Kant's that heavily emphasizes that morality is grounded in rationality (free thought, free will, and free action) will not see much importance in the evolutionary origins of humanity. Nagel states that ethics governs action, not just beliefs, and so ethics is concerned with motivation (Nagel 1978: 198). But because ethics is most importantly about reasoning, thinks Nagel, biology has virtually no value for ethical theory.

Like Nagel, who is aware of attempts to put ethics on a biological foundation but is unmoved by them, Korsgaard, too, is aware of a similar project: the temptation "to see the origins of morality in animal behavior" (Korsgaard 2006: 106). As we have seen, she claims that ethics arises from autonomy; and she also holds that "how far autonomy extends in the animal kingdom is an empirical question" (Korsgaard 2006: 112). For Korsgaard:

> There is nothing unnatural, non-natural, or mystical about the capacity for normative self-government. What it requires is a certain form of self-consciousness, namely, consciousness of the grounds on which you propose to act *as grounds*. (Korsgaard 2006: 113)

Reflective distance is a form of self-consciousness and, importantly, "This form of self-consciousness is the source of reason, a capacity that is distinct from intelligence" (Korsgaard 2006:113). "Many animals are intelligent," Korsgaard concedes (Korsgaard 2006: 104). But intelligence and reason are two different things, she argues.

> Intelligence is the ability to learn about the world, to learn from experience, to make new connections of cause and effect, and put that knowledge to work in pursuing your ends. Reason, by contrast, looks inward and focuses on the connections between mental states and activities. (Korsgaard 2006: 113–14)

So, like Nagel, she sees that autonomy and reason are somehow a function of our organic structure, and that organic structure is a product of an evolutionary process, yet when reason and autonomy become possible in an organism, it "gives us a whole different way of being in the world" (Korsgaard 2006: 117). Therefore, an evolutionary perspective can contribute little if anything to normative ethical analysis.

9.3 MODELS OF EVOLUTION AND DEONTOLOGICAL ETHICS

9.3a Darwin and Deontological Ethics

Darwin was familiar with Kant's ethical theory.[6] In *The Descent of Man* Darwin quotes a few lines from Kant and engages with deontological ethics. He quotes Kant asking the question, "Duty! Wondrous thought . . . whence thy original?" (Darwin 1871: 70–71). It is a "great question" says Darwin, and he intends to answer it from the perspective of natural history. Darwin doesn't use the term "deontology" (neither does Kant); rather, Darwin calls this kind of ethic "formal morality" (Darwin 1874: 74). Darwin characterizes formal morality as the view that in order for an action to be legitimately regarded as a moral action, it cannot be "performed impulsively" but must be "done deliberately after a victory over opposing desires, or when prompted by some exalted motive" (Darwin 1871: 87). One of the main tenets of a formal morality is that a moral agent is bound to the authority of categorical moral imperatives, regardless of his or her desires.

Darwin argues against this view. In discussing ethics Darwin has frequently described actions in which "moral beings" are "impelled by an instinctive motive to help others" (Darwin 1871: 87–88). Although Darwin agrees that an agent who struggles against fear, for example, and yet performs a heroic action, will deserve more credit, Darwin does not think that struggling against opposing desires should be the criterion of a moral action (1871: 88). He gives several reasons for this, one of which points to the common occurrence that if we perform a certain action very often we will not need to deliberate or hesitate with that type of action any longer, it will become second nature to us. Darwin doesn't think that because we can now perform a particular action effortlessly due to repeated performance, we should refrain from calling it moral (1871: 88).

Rather than characterize moral actions as always opposed to desires, Darwin suggests that an action can be considered moral if it is performed by a being "who is capable of comparing his past and future actions or motives" (Darwin 1871: 88). An action performed by such a being is considered a moral action, he says, regardless of whether it is performed through instinct, habit, or opposition to inclination (Darwin 1871: 88–89).

Under Darwin's characterization, a moral being possesses "self-command," and "power of reasoning" (Darwin 1871, 86). As we have seen, Kant calls these characteristics of a moral being, autonomy (freedom) and rationality. Whereas Kant (and Korsgaard) gives an account of where duties come from simply in terms of the autonomy of a rational being—"I give myself the categorical imperative"—Darwin says that the impulse by which we are impelled to some of our best actions actually originates in our natural history, for

Darwin views morality as more about sympathy and social instincts. Kant's ethic emphasizes humans as rational beings, whereas Darwin emphasizes humans as social beings. Their background metaphysics differ as well.

Although Kantian-styled deontology may be overly rational and seem to exclude feelings of love, compassion, and happiness, it is possible to build a deontological theory out of non-rational building blocks in order to form a moderate deontology (Kagan 1998, chapter 3.2; Timmons 2008).

Darwin does acknowledge the importance of rationality for a full-bodied ethics, yet he emphasizes that other important elements of human nature must be in place first. He thinks that humans have the impulse to aid others, for example, but the improved ability to reason would guide them (Darwin 1871, 86). For Darwin, reason alone will not generate ethical standards such as the golden rule. In ethics, a human nature endowed with social instincts is more fundamental. In Darwin's view, the social instincts plus intellectual powers plus acquired habits lead to the foundation of morality—the golden rule (Darwin 1871: 106). Darwin's notion of a moral being and Kant's notion of a moral agent are roughly the same. There is nothing in Darwin's account of ethics that precludes the deontological principles of respect for autonomy or universality. Darwin explains that his account of ethics can approximate toward Kant's categorical imperative and notion of dignity. After the ethical elements are in place: "He might then declare . . . I am the supreme judge of my own conduct, and in the words of Kant, I will not in my own person violate the dignity of humanity" (Darwin 1871: 86).

For Darwin, then, it is not the case that when we take an evolutionary perspective on human nature the main elements of a deontological ethic become sterile. We can still: recognize our own autonomy as persons who have dignity, deliberate about our duties, and use the categorical imperative as a normative principle.

Between Darwin's and Kant's accounts of ethics two differences stand out, though: the explanation of where our duties come from, and the degree of their bindingness. As Ruse pointed out, the notion of absolute duties sits uncomfortably in any naturalistic ethic, not just one that views human beings as the product of evolution.

9.3b Dawkins and Deontological Ethics

Dawkins is familiar with Kantian deontological ethics (2006a: 224–32). And Dawkins acknowledges that Kantian ethics provides us with satisfying answers to various moral dilemmas (e.g., by using the categorical imperative that says rational beings should not be used merely as a means). Kant, Dawkins says, believed we have these kinds of moral responses because there are moral absolutes. But for Dawkins, we feel such moral intuitions because of our evolutionary heritage (2006a: 224–25). And Dawkins thinks that "morals do not have to be absolute" (Dawkins 2006a: 232).

Dawkins' fellow genic selectionist, E.O. Wilson, acknowledges that Kant was a great philosopher, but thinks Kant was wrong about human nature and morality. Wilson asserts that

> religious transcendentalism is bolstered by secular transcendentalism, with which it has fundamental similarities. Immanuel Kant, judged by history the greatest of secular philosophers, addressed moral reasoning very much as a theologian. (Wilson 1998: 248)

Wilson tries to sum up the key elements of Kantian ethics in a paragraph:

> Most important, and transcendental, *ought* has no place in nature. Nature, Kant said, is a system of cause and effect, while moral choice is a matter of free will, for which there is no cause and effect. In making moral choices, in rising above mere instinct, human beings transcend the realm of nature and enter a realm of freedom that belongs to them exclusively as rational creatures. (Wilson 1998: 248–49)

Like Dawkins, Wilson says Kant's ethic "has a comforting feel to it," but according to Wilson, "it makes no sense at all in terms of either material or imaginable entities, which is why Kant . . . is so hard to understand. . . . It does not accord, we know now, with the evidence of how the brain works" (Wilson 1998: 249).

Wilson seems to understand "transcendental" as a special metaphysical category. But "transcendent" need not be taken to be anything so extravagant. When Dennett, for example, points to our power to rebel against our genes (a power that genic selectionists like Dawkins and Wilson acknowledge), Dennett says: "[T]hat is our transcendence, our capacity to 'rebel against the tyranny of the selfish replicators,' as Dawkins says, and there is nothing anti-Darwinian or antiscientific about it" (Dennett 1995: 471).

Wilson acknowledges this special power to rebel and even locates it in our intellect. A few pages after critiquing Kant, Wilson defends his own view of ethics and he describes how human beings have instincts that need to be molded by ethical principles.

> The essential ingredient for the molding of the instincts during genetic evolution in any species is intelligence high enough to judge and manipulate the tension generated by the dynamism. That level of intelligence allows the building of complex mental scenarios well into the future. . . . It occurs, so far as known, only in human beings and perhaps their closest relatives among the higher apes. (Wilson 1998: 251–52)

Wilson may feel that Kant's language is too "theological sounding," and Wilson may experience unfounded worries about the phrases "realm of nature" and "realm of freedom," but a Kantian makes the case that it is simply our unique intellect that allows us to transcend our instincts and use "mental scenarios" to formulate moral judgments and make ethical decisions. With Dawkins' model of evolution, it is granted that humans can exert a kind of transcendence over biological instincts. "The brain is big enough to override the genes," writes Dawkins (2001: 11). And this is tantamount to the autonomy that deontologists see as the basis of normative ethics.

Psychological attributes are put in place by evolution, says Dawkins. "Kin selection and selection in favour of reciprocal altruism may have acted on human genes to produce many of our basic psychological attributes and tendencies" (Dawkins 1989: 191). Even though Dawkins is a genic selectionist, he still acknowledges that humans are individual subjects. "Bodies may be colonies of genes," says Dawkins, "but they have an individuality of their own. An animal moves as a coordinated whole; subjectively I feel like a unit, not a colony" (1989: 46).

In terms of moral sensibilities, Dawkins says that moral sensibilities are simply built into human beings: "Sensitivity for unfairness is built into us (e.g., how I feel when I pay taxes but others do not)" (Dawkins 2001: 10). Kantian deontologists are simply offering a rational defense of these sensibilities—by using the concept of universalizability and the principle of the categorical imperative. And who are those beings who state that they wish to be treated fairly? Persons. "Persons" and "rights" may be looked at as ethical concepts for articulating what Dawkins regards as inbuilt moral sensibilities. (Critics will of course regard any attempted *rational* defense of moral sensibilities as rationalizing).[7]

Although it is true that moral absolutes are not "derived" from an evolutionist perspective, yet if a human intellect perceives the importance of moral absolutes in moral decision making, one need not reject moral absolutes because they are not derived from evolutionary considerations. If so, that would be what Nagel described as analogous to "abandoning a mathematical belief on other than mathematical grounds"; it would be abandoning a moral belief on biological grounds. Nagel regards this kind of maneuver as "irrational" (Nagel 1997: 244). A normative ethic is an ethic that can be used to make decisions. Dawkins takes a genic selectionist evolutionary perspective, but still acknowledges that we can make decisions based on values. He says:

> The brain has been equipped by the natural selection of genes with the power to take (sic) its own decisions—decisions based not directly upon the ultimate Darwinian value of gene survival, but upon other more proximal values, such as hedonistic pleasure or something more noble. (Dawkins 2001: 11)

As noted above, Dawkins and other genic selectionists accept that humans can transcend their genes:

> It is a manifest fact that the brain—especially the human brain—is well able to over-ride its ultimate programming; well able to dispense with the ultimate value of gene survival and substitute other values. I have used hedonistic pleasure as just an example, but I could also mention more noble values, like a love of poetry, or music, and of course the long-term survival of the planet—and sustainability. (Dawkins 2001: 11)

Or the absolute value of human persons. It is rational to believe we have duties to preserve and protect things that have value (which could include persons whom we regard as having dignity).

9.3c Gould and Deontological Ethics

Dawkins himself ultimately finds Kantian ethics unsatisfactory, but it is not because genic selectionism cannot accommodate the elements of deontological ethics. The same is true of Gould and his model of evolution: although he himself has reservations about Kantian ethical principles, Gould's model of evolution still allows for the building blocks of a deontological ethic.

Writes Gould:

> I have never been much attracted to the Kantian categorical imperative in searching for an ethic—to moral laws that are absolute and unconditional, and do not involve any ulterior motive or end. The world is too complex and sloppy for such uncompromising attitudes. . . . I prefer the messier "hypothetical imperatives" that invoke desire, negotiation, and reciprocity. Of these "lesser," but altogether wiser and deeper principles, one has stood out for its independent derivation, with different words but to the same effect, in culture after culture. . . . Christians call this principle the "golden rule"; Plato, Hillel, and Confucius knew the same maxim by other names. I cannot think of a better principle based on enlightened self-interest. (Gould 1993: 50)

Here Gould writes about ethical approaches in terms of his attraction and preference for them. His point against Kantian deontological ethics is that it doesn't fit with a complex and sloppy world. Maybe so. But should it? Kant's point about ethics being about the realm of freedom is because ethics is meant to free us from the world around us.

And Gould admits just as much as the genic selectionists do, that humans have a power to rebel against their genetic nature. Gould is very confident that we need not assume that there is a genetic basis to our most fundamental traits, the "predispositions of our genetic constitution" (Gould 1977: 238).

This means that humans can make themselves into beings other than what we might think if we only focused on biology.

9.3d Haught and Deontological Ethics

E.O. Wilson has claimed that Kant's language *sounds* theological. But the elements of Kant's ethical theory—duty (obligation), autonomy, law, dignity, the categorical imperative, person, rights, universalizability, and nonconsequentialism—need not be framed theistically.[8] From what we have seen in previous chapters, Haught, by contrast, obviously *does* frame ethics theistically. And so Haught would say that Kant's deontological ethics, actually, along with all other ethical theories too, will fall short if it lacks a theistic foundation.

Haught senses no difficulty in coupling a deontological ethic with a theistic model of evolution. In fact, Haught believes that an evolutionary perspective can shed important light on ethics. The central evolutionary contribution to this fresh perspective comes from "our new evolutionary awareness that we live in an unfinished universe" (Haught 2003: 49). The awareness that we live in an unfinished universe has interesting implications for ethics, for how we should live. In the light of this perspective, says Haught, we should act in ways that leave open "incalculable outcomes," since creation is still unfinished as it moves into the future (Haught 2000: 159).

One argument that Kant uses to justify a nonconsequentialist ethic is that the alternative—a consequentialist ethic—presumes that human beings can have knowledge of outcomes and use that knowledge to determine if actions are morally right or wrong. Kant says we cannot have that kind of knowledge in our ethical decision making; we must be content with using our will (and the categorical imperative) as the barometer for ethics. Further, Kant believes, we have control over our own will, yet we do not have control over the ultimate outcomes of our actions. Haught's theistic evolution, in short, which takes seriously the notion of "incalculable outcomes," fits well with a nonconsequentialist, deontological ethic.

Haught actually mentions Kantian ethics explicitly (2006; 2010). Haught specifically says he is not using a Kantian strategy, but that Kant had "the correct intuition that being responsible means *being grasped* by a dimension of reality that radically transcends nature and the human subject" (Haught 2006: 155). There are several Kantian elements that Haught actually does employ: nonconsequentialist elements (as noted above), imperatives, and the importance of acknowledging something as good in itself.

In chapter 2 (§2.3d), we saw how Haught sums up the roots of ethics in the imperative to "be responsible," which is part of human critical intelligence. Notice how the very notion of an imperative of critical intelligence is similar to the notion that the categorical imperative is a command of our reason. Haught's account differs from Kant's, though, because Haught does not hold that reason itself is sufficient to generate the moral imperative. For Haught, the imperative to be responsible derives from the human mind's anticipation of transcendent goodness (Haught 2006: 155).

And here we see how Haught does indeed use the term "transcendent" to mean that the natural world will not sufficiently explain the source of the moral imperative, because the ultimate source of the moral imperative is "a transcendent goodness that encompasses and grounds" the world. It sounds like Haught is referring to God; and he is. Another similarity with Kant is the reference to something that is good in itself. Kant uses the concept of a good will, a will as something good in itself. In Kant's approach, however, there need be nothing supernaturalist about this goodness. Haught, though, pushes the notion of goodness in itself to its extreme and, as with traditional theism, says that when we are talking about goodness itself we are really talking about God.

A concrete example of how Haught takes transcendent goodness as evidence that naturalistic evolutionary accounts of ethics will fall short is Haught's reference to "a rare stage of moral and religious development" (Haught 2006: 161). This stage of moral development is achieved by a few heroic individuals who have not feared to bring disharmony to their conventional settings (Haught 2006: 162). About these rare individuals Haught also says that they exhibit

> the kind of conduct shaped by the conviction that certain norms and actions are intrinsically "good" and that one is obliged to pursue them no matter what cost to oneself, one's group, or the genes of one's relatives. (Haught 2010: 121)

These are "exceptional human beings" that cannot be explained in naturalistic evolutionary terms. According to Haught: "Those heroic, prophetic, individuals are attracted to something, having been grasped by something imperishably and absolutely good" (Haught 2006: 162). An ethic that says there are actions and norms that one is *obliged* to pursue no matter what the consequences is known as a deontological ethic.

A sticking point between an evolutionary perspective and deontological ethics that has come up with several authors in this chapter is deontology's absolutism. Haught's model of evolution provides an interesting response to this issue, and the response is derived from his distinctively evolutionary perspective. In Haught's evolutionary theory, an unfolding universe is not *directed* to its end, because God, as love, does not force nature to be what it is; but as with those rare moral individuals, nature is allowed to evolve and develop in its own way, only being attracted to a transcendent goodness. We are allured to the good in ways we may not even be able to articulate. This makes sense, says Haught, because it is in the very nature of love to not coercively manipulate people, so we should not expect a God of love to force nature to perfection (Haught 2000: 53).

God works by "divine allurement" and "an enticing and attracting divine humility" (Haught 2000: 53). Even though we have the feeling of obligation when we become aware of intrinsic goods (and that obligation can be rationally justified), yet because love does not coerce, Haught's ethic that he supports with evolutionary considerations is not a rigidly absolutist deontological ethics.

9.4 CONCLUSION

Kantian deontological ethics takes commonsense ethical notions and justifies them with various abstract concepts. While "normative self-government" and "autonomy" may sound abstract, yet we have seen that evolutionists from Darwin, to Dawkins, to Wilson, et al., acknowledge that this is an intellectual ability that human beings possess. In this chapter we have seen that autonomy and other key elements of deontological ethics can make practical sense even when we take an evolutionary perspective. Traditional deontological ethics has usually been framed in a theistic metaphysics, so it was no surprise that Haught was able to assimilate a deontological ethic into his theistic evolutionist perspective.

While some evolutionists, such as Wilson, Dawkins, and Gould have argued against Kantian deontological ethics, we have seen that neither an individualist, genic selectionist nor a multilevel selectionist evolutionary view undermines the basic building blocks of a deontological ethic. Deontological ethics, we must remember, is a normative ethics. An easy way to avoid mistakes is to think of Gould's point about how the categorical imperative does not align with a complex and sloppy world. The fact that a normative ethical principle advises what ought to be done, as opposed to advising to continue what *is* done in the natural world, is not thereby a mark against that normative principle. Normative principles are *meant* to do more than simply describe and align with the world as it is.

E.O. Wilson, similarly, has the mistaken notion that ethics is a discipline that is meant to explain the reasons why people do things. Wilson says:

> A science of sociobiology . . . might transform the insights of ancient religions into a precise account of the evolutionary origin of ethics and hence explain the reasons why we make certain moral choices instead of others at particular times. (Wilson 1980: 63)

Investigating the reasons why people choose to do things is an interesting and worthy enterprise, but such a project on its own does not give us normative ethical principles. Normative ethics has a different goal: it inquires into how we *should* make choices and what makes a choice or an action *good*.

Once we are clear about what kind of theory we are investigating here—a normative one—we can recognize that even Dawkins acknowledges that we can choose the values we wish to use to guide our moral decision making and our life choices.

Nagel and Korsgaard, as defenders of deontological ethics, have stated that biology and evolution do not contribute to normative ethical analysis. A claim such as that simply demarcates normative inquiry from biological inquiry. And that is true as far as it goes. Yet the question still remains as to whether an evolutionary perspective will undermine a deontological ethics.

When we look at the building blocks of deontological ethics, especially autonomy—the most significant feature of morality according to deontologists—we have seen that evolution can provide the raw materials, but it is still up to human beings to decide and choose how they should act. Kant points to that power when he calls rational beings "legislative beings." Rational agents have the capacity, as Korsgaard says, to assess and then adopt their intentions. Interestingly, the utilitarian ethical perspective we will examine in the next chapter downplays the importance of intentions in ethics. Utilitarian ethics recommends that when determining the right thing to do, we should be most attentive to the outcomes and consequences of our actions. Attempts to join utilitarian ethics with an evolutionary perspective go all the way back to the nineteenth century (§1.4).

NOTES

1. Regarding autonomy, Munro, who has extensively studied Chinese philosophy, says that autonomy has only recently emerged as an important value in China; this was due to historical differences between the East and the West, he claims (2005: 39–41).

2. For example, to be a Kantian deontologist one need not endorse the particular metaphysical system that Kant has devised, which countenances a *noumenal* realm to reality in addition to a *phenomenal* realm (a realm of things-in-themselves as contrasted with a realm of observable objects).

3. In Kant's writings, another interesting connection to evolution is his argument about nature, happiness, reason, and instinct. In the natural constitution of an organized being, he says, no organ is to be found for any end unless it be the most fit and adapted for that end. If nature wanted us to achieve happiness, then it would have done it most efficiently by implanting in us an instinct to achieve that end. But nature has not done so. We do have reason, though. Could the purpose of our reason be to guide us to achieving happiness? No, argues Kant. We can see that reason does a poor

job in bringing us happiness—the more it tries, the more it fails. Kant concludes that nature has not imparted reason to us for the purpose of happiness, but for the purpose of producing a will that is good in itself (1785: 8–9).

4. Social contract ethics and utilitarian ethics can be regarded as consequentialist ethical theories; and so can be regarded as adopting "hypothetical imperatives." Hobbesian social contract ethics, for example, can be summarized as offering the following as a moral imperative: "If you want to protect your self-interest, then you should leave the state of nature and adopt a social contract." Utilitarian Ethics (as described in chapter 10), can be summarized as offering the following as a moral imperative: "If you want bring about the greatest happiness for the greatest number, then you should tell the truth, keep your promises, don't harm others, etc." Precisely because these are *hypothetical* imperatives, Kant says they are not *moral* imperatives; for Kant, only *categorical* imperatives are moral imperatives.

5. There may be some passages in Kant, where, caught up in enthusiasm he may blur the distinction between offering an account of how humans ordinarily make moral judgments, and how humans *should* make moral judgments.

6. The writer and activist Frances Power Cobbe met Darwin and they corresponded. Cobbe was a defender of Kantian ethics and she mailed Darwin a work by Kant. As a Kantian, she disagreed with Darwin's characterization of ethics. See Cobbe 1871.

7. See Greene (2008).

8. Kain (2005) reviews the theological dimensions of Kant's ethics.

Chapter 10

Evolution and Utilitarian Ethics

In chapter 8 on evolution and social contract ethics I mentioned that evolutionists often incorporate elements of social contract thought, sometimes implicitly and sometimes explicitly. In this chapter on utilitarian ethics and evolution we come across the same situation. E.O. Wilson, for instance, who often comments on the poverty of moral philosophy, then takes it upon himself to formulate an "empiricist view" of ethics that sees ethics as based on "*strong innate feelings*" that "*cause certain actions to be preferred*" since "*we have experienced them, and weighed their consequences*" (Wilson 1998: 251, italics in original). In this case, Wilson the evolutionist characterizes ethics in rather standard utilitarian fashion, seemingly without even realizing it.

In this chapter I will first lay out the main points of utilitarian ethics that are pertinent to the discussion, concepts such as sentiment, sentience, happiness, consequentialism, altruism, and egalitarianism; and principles such as the principle of utility and the principle of equality. Following that I will look at what different writers (Rachels, Ruse, Dennett, and Singer) have written about utilitarian ethics and evolution. And finally I will examine what contemporary models of evolution imply about utilitarian ethics.

10.1 UTILITARIAN ETHICS

Traditional utilitarians argue that ethics is based on feelings. David Hume argues for this general thesis, and his well-known argument for such an ethical framework contains two premises (Hume 1739, Book 3, part 1, sect. 1). The first premise is that ethics concerns how we ought to live and what we should do; and the second is that human reasoning is not the kind of thing, or kind of capacity that can motivate us into action, since reason, he says, is powerless for producing or preventing actions. Hume then reaches the conclusion that ethics cannot be based on human reason and rationality, but must be based on sentiments. Hume's argument, in short, is that if ethics has to do with actions and guiding our actions, then ethics must have to do with feelings and emotions, because sentiments are the only things that can really motivate us into action. Moreover, in order to make an ethical judgment, one needs to feel something; for again, reason alone is impotent (Hume 1739: 413–15).[1] The word *sentiment*, from the Latin word *sentire*, *to feel*, will prove to be highly significant for the utilitarian tradition: *sentience*, which comes from the same Latin root, is an important concept for utilitarians. The classical utilitarians, Bentham and Mill, both put feeling at the foundation of their hedonistic utilitarianism. Bentham asserts that "nature has placed mankind under the governance of two sovereign masters, *pain* and *pleasure*" (Bentham 1789: 1); and Mill contends that the ultimate sanction of all morality is a subjective feeling in our own minds (Mill 1861: 28).[2] But how do we move from a theory that says morality is based on feelings, to a "utilitarian" normative ethic?

The word *utility* just means "usefulness," so the question is: "useful for what?" What is a "useful" ethics? For a utilitarian, the answer is uncomplicated: a useful ethic is one that is beneficial, one that brings benefits; it brings about things that people desire and that will benefit them. For Bentham, the most basic way to categorize our feelings is as good feelings or bad feelings. But what makes all the good feelings, good? For Bentham, the issue is straightforward: good feelings are pleasurable. The experience of pleasure makes good feelings good, and the experience of pain makes bad feelings bad. According to utilitarians, people desire—have strong positive feelings for—happiness. And people have strong negative feelings toward pain, suffering, and unhappiness. As Mill describes it,

> The creed which accepts as the foundation of morals "utility" or the "greatest happiness principle" holds that actions are right in proportion as they tend to promote happiness; wrong as they tend to produce the reverse of happiness. By happiness is intended pleasure and the absence of pain; by unhappiness, pain, and the privation of pleasure. (Mill 1861: 7)

For the utilitarian, individuals want these positive things—pleasure, good feelings, happiness—for themselves, and they want to avoid pain, displeasure, and unhappiness. And when an individual sees *others* in pain, it is possible for that individual to feel the other person's pain; this is literally what happens with *sympathy*. Part of the utilitarian view of human nature is that human beings have natural feelings of sympathy (*sym* "together" + *pathy* "feeling") for other human beings. Instead of saying that humans are natural-born egoists, utilitarians hold that humans are born with the potential for altruism. The utilitarian view of human nature is that human beings have genuine feelings for others (in addition to feelings for themselves, of course).

But *why* do we feel others' pain and suffering when it is not actually happening to us? If psychological egoism were true and human beings were selfish through and through, it wouldn't matter to us what happens to other people, especially other people we don't even know and never have even met. The utilitarians reject psychological egoism, though.

In utilitarian thinking, as soon as we care about what is happening to others, we then begin to analyze events and actions in terms of how they contribute to the pleasure or suffering of other people. We care about the consequences of our actions on others. We consider whether our action will contribute to their good feelings; we consider whether our action will contribute to their suffering. In other words, we think about the usefulness, or utility, of a particular action on our feelings and on others' feelings. With utilitarian ethics I judge a rule or law to be right or wrong depending on the consequences (or utility) that it will have not just on me (that would be an egoistic ethics), but on *all* who will be affected. Unsurprisingly, utilitarian ethics is a *consequentialist* ethic: in deciding when an action, rule, policy, law, or practice is ethical, utilitarian ethics focuses on outcomes, consequences. We care about what consequences our actions will have both for us and for others. According to Mill, "the happiness which forms the utilitarian standard of what is right in conduct is not the agent's own happiness but that of all concerned. As between his own happiness and that of others, utilitarianism requires him to be as strictly impartial as a disinterested and benevolent spectator" (Mill 1861: 16). This notion prompts Mill to say that the golden rule sums up the spirit of utilitarian ethics. "In the golden rule of Jesus of Nazareth," Mill writes, "we read the complete spirit of the ethics of utility. 'To do as you would be done by,' and 'to love your neighbour as yourself,' constitute the ideal perfection of utilitarian morality" (Mill 1861: 16–17).

The argument that moves from saying that we *do* have these feelings for others, to the normative claim that we *should* gauge our actions based on how they impact others in addition to ourselves, is principally an argument of consistency. To go from a claim about our human nature (an *is* claim) to a *moral* claim (an *ought* claim) that we *should* do this, and it is *right* that we do this, and *wrong* when we do not do this, involves an important step in the argument. The crucial step is to ask ourselves whether there is really a difference between our joys and pains and others' joys and pains. If there really is no difference between our pleasures and pains and others' pleasures and pains then we *should*, just because of consistency, regard their suffering as just as important as ours. This is the core of the justification of the principle of utility—we ought to do whatever will have the best consequences for all concerned, not just for ourselves, because there really is no important difference between our welfare and others' welfare. It is an argument that goes from saying we *do* do this, that's what humans are like, to the *moral* claim that we *should* do this. At the heart of the main utilitarian argument that moves from the concern we naturally have for our own feelings of pleasure and pain, to others' feelings of pleasure and pain, is the belief that that's what humans are like. When we hear about tragedies befalling others we may find ourselves cringing or grimacing. In those cases we genuinely feel for others, we feel their joys and their pains. We feel our own joys and pains too. So the question is: Is there a difference between our joys and pains and their joys and pains? If there is no difference, then we *should*, just because of consistency, regard their suffering as just as important as ours. Thus, the principle of utility—one *ought* always to do whatever will have the most utility for *all concerned*—is the main utilitarian principle. Utilitarianism is an ethical approach that focuses on happiness, not only on the individual's happiness, but everyone's happiness. If there is nothing different about others' pain and suffering and our own pain and suffering, then, to be consistent, we should be concerned about the consequences for all concerned. Although utilitarian ethics, going all the way back to Hume, rests on sentimentalist foundations, there are, however, some aspects of utilitarian ethics that do not solely depend on sentiments. The consistency argument above, for one. Further, the hedonistic calculus, or any other kind of utilitarian cost/benefit analysis, depends on logical or quasi-mathematical measurements. This is consistent with Hume's dictum that "reason is the slave of the passions" (Hume 1739: 415). For although we are driven by our sentiments and desires, there is still a place for human reason in ethics.

If we reflect on the notion that our happiness is not much different from others' happiness and we should regard others' happiness on a par with ours, then we have backed into another important ethical principle, the principle of equality: "The interests of every being affected by an action are to be taken into account and given the same weight as the like interests of any other being" (Singer 1975: 5). It is clear that equality is a key concept that is implied in this reasoning. For, another way to describe what we have been talking about and capture the central utilitarian idea, is simply to say that humans are equal: your suffering or happiness is equal to my suffering or happiness. My happiness, suffering, well-being, pleasure, and pain, are not more important than yours. Seeing ethics along utilitarian lines brings us from egoism to altruism to equality.[3]

In summary, the pertinent utilitarian ethical concepts are: sentiment, sentience, happiness, altruism, consequentialism, and egalitarianism; and the key utilitarian principles are the principle of utility and the principle of equality.

10.2 UTILITARIAN ETHICS AND EVOLUTION

Michael Ruse asserts that "we may have selfish genes, but we are not necessarily always selfish humans" (Ruse 1991a: 64). In specifically discussing Hume's moral philosophy, Ruse says that "the Darwinian agrees that morality does basically come down to some kind of sympathy or fellow-feeling" (Ruse 1991a: 72). Ruse has considered whether utilitarianism as a normative ethic still makes sense in the light of Darwinian evolution. The willingness to help and cooperate that the Darwinian sees as the result of evolution, Ruse sees as consistent with utilitarian ethics (Ruse 1998a: 235). If we talk about sentience, Ruse agrees that "the Darwinian and the utilitarian are talking of the same types of thing—human moral sensitivities" (Ruse 1998a: 241). The Darwinian also agrees with the utilitarian's focus on utility, Ruse says, because our happiness is an important desired end (Ruse 1998a: 236). In Ruse's estimation, there is broad overlap between the Darwinian and the utilitarian view (Ruse 1998a: 237).

Ruse has some doubts about reconciling Darwinism with what utilitarian ethics says about the expansion of the moral community. To Ruse, the notion that human beings should consider the consequences of their actions for all who will be affected seems to push up against kin selection and reciprocal altruism. Biological feelings toward one's children will be stronger than feelings toward strangers on the other half of the world (Ruse 1998a: 237–39). Ruse is voicing the common objection that utilitarian ethics is too demanding. Darwin himself touches on this aspect of utilitarian ethics, but he does not mention that it is too demanding, he sees it as a noble characteristic. The virtue of humanity, he calls it, seems to arise from our sympathies getting more widely spread out to more and more sentient beings (Darwin 1871: 67).

Part of the utilitarian's response to the objection that it is too demanding is found in Darwin's notion that the extension of sympathies is due to an advanced intellect. Human beings are only in a position to see themselves in a broad moral community and recognize extensive obligations to aid those at the far reaches of the moral community when they have an ethics that has been developed at an intellectual level. It is at the intellectual level that the principles of utility and principles of equality acquire their normative force because it is only at that level that human beings can ask themselves whether their own pleasure/happiness is more important than others' happiness. Rational consistency nudges such reflective beings to realize that their individual happiness is *not* more important than others' happiness so they therefore *ought* to perform those actions that will bring about the best consequences for all concerned, and they *ought* to regard others' interests as on a par with their own. Some have given the impression that if human biological evolution is part of our worldview then commonplace notions of ethics no longer fit and need to be abandoned so that we can create ethics anew. In the article about moral philosophy that Ruse coauthored with E.O. Wilson (1986), for instance, they give the impression that recent evolutionary theories have ushered in a new way of understanding ethics. But the standard utilitarian starting point appears in their actual claims about ethics. They say that ethics "is rooted in contingent human nature, through and through" (Ruse & Wilson 1986: 186). Utilitarian ethics does not assume that there are "extrasomatic guides" that underwrite "genuine objective external ethical premises," as they claim, so it is not plausible for Ruse and Wilson to portray utilitarianism as the kind of moral philosophy that has been eclipsed by the modern synthesis of evolutionary theory (Ruse & Wilson 1986: 186). As mentioned earlier, utilitarian ethics is naturalistic about ethics, not supernaturalistic.

According to Wilson: "Human emotional responses and the more general ethical practices based on them have been programmed to a substantial degree by natural selection over thousands of generations" (Wilson 1978: 6). Because of these origins he believes a new morality will have to be forged by new ethicists (Wilson 1978: 196). But Wilson's "search for a new morality" seems unnecessary (Wilson 1978: 4). For, again, he accedes to the basic building blocks of a utilitarian ethic. What Wilson calls his "empiricist principle of ethics" bears the marks of utilitarian ethics. In the following passage that sums up his empiricist principle of ethics he adverts to the utilitarian concepts of moral sentiments and consequences: "*Strong innate feelings and historical experience cause certain actions to be preferred; we have experienced them, and weighed their consequences, and agree to conform with codes that express them*" (Wilson 1998: 251, italics in original). Evolution by natural selection has not prompted Wilson to reject a utilitarian ethic; on the contrary, Wilson's biological view has prompted him to view ethics in rather standard utilitarian fashion. He seems to be presupposing a consequentialist ethical framework, because he insists that ethics depends on the accurate prediction of consequences of our actions (Wilson 1998: 240). In a later work, Wilson does admit that his ethical approach has roots that go back to Aristotle, Hume, and Smith (1998: 248, 251).

In chapter 17 of Dennett's *Darwin's Dangerous Idea* (1995), which is entitled "Redesigning Morality," Dennett also gives the impression that in the light of Darwinism we will need to create ethics anew. In that chapter he specifically critiques utilitarian ethics. But the criticisms he wages against utilitarian ethics have been offered many times before and they do not derive from a uniquely Darwinian perspective. In a later work (2003, chapter 7) Dennett discusses—somewhat optimistically—the varieties of altruism that are possible under a Darwinian framework. His "Moral first aid manual" (1988) aspires to be more practical and less theoretically idealistic as compared with most ethical theories; but in no way does he show that utilitarian ethics is in need of revision or abandonment because the view of human nature it relies upon is inconsistent with a human nature that has been formed through thousands of generations of genic selection.

For Dennett, Darwin's dangerous idea is dangerous for ethics because:

> There is no denying, at this point, that Darwin's idea is a universal solvent, capable of cutting right to the heart of everything in sight. The question is: what does it leave behind? I have tried to show that once it passes through everything, we are left with stronger, sounder versions of our most important ideas. Some of the traditional details perish, and some of these are losses to be regretted, but good riddance to the rest of them. What remains is more than enough to build on. (Dennett 1995: 521)

I would say that these statements can apply to ethics, and that there is a considerable amount that remains, even after Darwin's dangerous idea has cut through. And in a later work, Dennett acknowledges this point about ethics. "It is commonly thought," writes Dennett, "especially by those who fear any invocation of evolutionary considerations in ethics—that since Darwinism sees 'nature red in tooth and claw,' it can only subvert or discredit our ethical aspirations, never support them with new insights, new foundations. This is simply not true" (Dennett 2003: 193).

Peter Singer, having examined evolutionists' claims about human nature, believes that utilitarian ethics still fits in and makes good sense even in the light of evolutionary considerations (1981; 1999). Against the backdrop of evolution by natural selection, the utilitarian's ethical principle of equal consideration of interests, he says, still maintains its normative force (1981: 109–10).[4] For Singer, in the face of human biological evolution, we need not abandon traditional utilitarian ethics nor do we need to create ethics anew. Singer does say, though, that certain ethical principles and theories (such as nonconsequentialist ethical theories that don't define right or wrong in terms of consequences) may seem less plausible through an evolutionary perspective (1981: 67). And in the light of a Darwinian view of human nature, social/political reformers who rely too heavily on the belief that human nature is malleable need to rethink their assumptions (1999: 24). But utilitarianism, though, as a kind of consequentialist ethic is perfectly comfortable in incorporating new information as it becomes available, whether through the sciences, or other avenues (Singer 1981: 64). For Singer, a utilitarian ethical position is actually bolstered by Darwinian views on human nature; for "an understanding of human nature in the light of evolutionary theory can help us to identify the means by which we may achieve some of our social and political goals, including various ideas of equality, as well as the possible costs and benefits of doing so" (Singer 1999: 15).

10.3 MODELS OF EVOLUTION AND UTILITARIAN ETHICS

10.3a Darwin and Utilitarian Ethics

James Rachels has noted that Darwin was able to maintain a vision of human nature as having been created through the process of natural selection, while at the same time maintaining a utilitarian moral vision (1990: 158–58, 223).

In *The Descent of Man* Darwin touches on most of the elements of utilitarian ethics. As we have noted, Darwin quotes both Adam Smith (Darwin 1871: 81, no. 17) and David Hume (Darwin 1871: 85, no. 19) on sympathy. Darwin seems to concur with Hume that ethics is ultimately based on feelings. And in utilitarian fashion, Darwin acknowledges and emphasizes that human beings have feelings not only for themselves, but for other human beings, most obviously the "parental and filial affections" (Darwin 1871: 41). Darwin observes that there are social instincts in animals and humans (1871: 63–64). He recommends that we "take as the standard of morality, the general good or welfare of the community" because the social instincts in animals and humans have likely developed for this reason (Darwin 1871: 63–64).

Darwin seems to attribute utilitarian reasoning to early humans when he says that actions were probably regarded as good or bad depending on how they affected the tribe (Darwin 1871: 63). To say that actions are good or bad as they affect the welfare of the tribe is a consequentialist ethic. Darwin further says that our valuing of others does depend both on our feelings for others as well as the ability to reason out the distant consequences of our actions (Darwin 1871: 58).

For the utilitarian, you have an innate feeling of sympathy for others and when you couple that with your capacity for reasoning that others would be better off if you helped them than if you didn't help them that translates into an ethical obligation to help them. Darwin's recipe for a moral sense is that any animal with well-developed social instincts would acquire a conscience (moral sense) as soon as the intellectual powers such as we see in humans are in place (Darwin 1871: 41–42).

With regard to happiness, Mill says a utilitarian standard requires that we treat others' happiness as on a par with our own, which he says is the essence of the golden rule. Darwin, too, as we've seen, views his ethical view as consistent with the golden rule (Darwin 1871: 71). From the utilitarian perspective, it is only a short step from here to egalitarianism and the principle of equality, and these ethical elements interlock with Darwin's discussion of extending feelings of sympathy to other beings beyond one's tribe, and his acceptance of a sentiment-based ethical theory.

For Darwin, what presumably prevents early humans from considering the welfare of other tribes, the whole species, or other species for that matter, is their limited capacity for reasoning. Because as humans developed the intellectual abilities to allow them to foresee the distant consequences of their actions, then their sympathies could be extended beyond the tribe, and the race, and even to other animals (Darwin 1871: 69). The very social instincts themselves, on Darwin's lights, are geared toward bringing about the best consequences for the community. All human beings must do is allow themselves to feel those sympathies and be aware of the far-reaching consequences of their actions on the well-being of others. The other name for the principle of utility—the greatest happiness principle—Darwin names as the standard (not as the motive) of conduct (1871: 64).

Hedonism—the view that the good is to be defined in terms of pleasure—was incorporated into the early versions of utilitarianism. Darwin, in his assessment of utilitarian ethics, sees hedonism as a flaw in the theory. For, in his view, it is simply not true that pleasure or displeasure is associated with each human action. There are plenty of examples, Darwin writes, where human beings experience an impulsive power that is different from a search after pleasure or happiness (1871: 64). Darwin envisions many cases in which an individual may make a sacrifice for the general welfare of the community. Since these are instances of sacrifice of pleasure and happiness for the individual, Darwin takes these as marks against utilitarianism, the greatest happiness theory.[5] Like consequentialist ethical theorists today, Darwin wishes to distinguish his ethical view from "utilitarianism" because utilitarianism comes prepackaged with hedonistic components. But as with consequentialists today who accept a sentiment-based ethic, Darwin's view does seem quite close to utilitarianism broadly understood, with the hedonistic component excised out.

By Darwin's own treatment, it seems that most traditional utilitarian ethical concepts and principles make sense even when placed in an evolutionary framework.

10.3b Dawkins and Utilitarian Ethics

As a genic selectionist, Dawkins will regard social instincts as the properties of vehicles; and those vehicles have the properties they do because they have (statistically) aided their genes in persisting. Behaviors are complex adaptations to aid the preservation of, ultimately, genetic replicators. Dawkins uses examples such as beaver dams, spider webs, and caddis houses, organisms that have developed elaborate repertoires of behavior that manipulate the world around them. As Dawkins says: "The replicators that exist tend to be the ones that are good at manipulating the world to their own advantage" (Dawkins 1982: 264).

For a genic selectionist like Dawkins, sentiment and sentience are human phenomena that have a deeper genetic explanation. Genic selection gives a particular account of *why* humans have social instincts. But a genic selectionist does not deny that humans can indeed *possess* social instincts or social sentiments. To get a traditional utilitarian ethic off the ground, we need to admit that human beings have natural (instinctive) feelings for at least some other human beings. Genic selectionists do not deny that human beings care about the suffering of others. They suggest we bear in mind, though, that caring for others' suffering has to do with the survival of the carer's DNA. According to Dawkins, "DNA neither knows nor cares. DNA just is. And we dance to its music" and "Nature is neither kind nor unkind. She is neither against suffering nor for it. Nature is not interested one way or the other in suffering, unless it affects the survival of DNA" (Dawkins 1995: 131, 133).

Dawkins' phrase "selfish gene" is often misunderstood in popular culture to mean that he denies the existence of social sentiments or social instincts, or he is claiming that human beings ought to be selfish.[6] But Dawkins himself says that he is not advocating selfishness as a principle by which we should live (1989: 267). Associating selfishness with evolutionary thinking did not have its origin with Dawkins, though. In the popular mind, Darwinism is constantly associated with competition, and to be competitive one must seek to advance one's own interests, not competitors' interests.[7]

For utilitarians, human beings aim for happiness, and they should aim for happiness. Dawkins maintains that evolution cares not for human happiness.

> [W]hen . . . that which is being maximized—is DNA survival, this is not a recipe for happiness. So long as DNA is passed on, it does not matter who or what gets hurt in the process. . . . Genes don't care about suffering, because they don't care about anything. (Dawkins 1995: 131)

And evolution doesn't care about *group* happiness either.

> Humans have a rather endearing tendency to assume that welfare means group welfare, that "good" means the good of society, the future well-being of the species or even of the ecosystem . . . a contemplation of the nuts and bolts of natural selection, turns out to be sadly at odds with such utopian visions. To be sure, there are occasions when genes may maximize their selfish welfare at their level, by programming unselfish cooperation, or even self-sacrifice, by the organism at its level. But group welfare is always a fortuitous consequence, not a primary drive. This is the meaning of "the selfish gene." (Dawkins 1995: 121–22)

Our genes may not care for individual or group happiness or the absence of suffering, but we as psychological beings with social sentiments surely do, and as psychological agents we must still decide what we ought to do.

We can look to these psychological capacities for underwriting the utilitarian's two main ethical principles: the principle of utility and the principle of equality. The principle of utility says that we ought to do what will bring about the most happiness for all concerned. But why should we? The utilitarian says it is because we want happiness for ourselves; and our happiness, intimately linked with our ability to experience pleasure and pain, is not more important than the happiness of others who can have the same kinds of experiences. Thus, we *ought* to look at our actions and consider what effect they will have, not only on our happiness, but on the happiness of all who will be affected by our actions.

In the nuts and bolts of natural selection as Dawkins understands it, we find genes striving to maximize their own welfare; we do not find a utopian vision of genes striving for the future well-being of the species or of a group, or ecosystem. But even if it is true that at the genic level genes are striving for their own selfish maximization, a utilitarian ethical principle still recommends that we as beings with developed psychological capacities (i.e., psychological and moral agents), in order to be consistent about our own striving for happiness, ought also to strive to bring about the greatest happiness for the greatest number. By acknowledging the presence of social sentiments and developed psychological capacities that allow human beings to conceptualize their own well-being and happiness as on a par with others' happiness and well-being, genic selectionism provides the necessary building blocks for a utilitarian ethic.

10.3c Gould and Utilitarian Ethics

Gould's hierarchical theory of selection has no difficulty incorporating the insights of kin selection to account for how affections for those who are closely related to us get put in place by natural selection. As Gould says,

> Our altruistic tendencies need not represent a unique overlay imposed by the demands of civilization. These tendencies may have arisen by the same Darwinian route via kin selection. (Gould 1977: 266)

Then again, though, according to Gould, whereas genetics may predispose human beings to various sorts of altruistic behavior, no specific behaviors are necessarily determined to follow from a particular genetic profile (Gould 1977: 266). For Gould, human culture and flexibility keep things open. "[S]o profound is our ignorance about the biology of human behavior," Gould claims (Gould 1977: 261).

Considerations such as these have bearing on the concept of utility. A utilitarian ethic regards something as right or wrong depending on how useful it is in contributing to our desired end, which in classical utilitarianism is human happiness. If we understand human nature as a product of evolution as Gould describes it, then what is our desired end? Human potential and human flexibility will have Gould conclude that there is no one single human end. Human beings act toward a plurality of ends. In Gould's view, my personal ends may even be religiously inclined, for he holds that there is no necessary conflict between science and religion if we understand religion, not science, as concerned with values and meaning. Since life has no direction and there is no ultimate purpose, Gould maintains that human beings must create meaningful lives for themselves.

Under Gould's model, an ethical system would seem to be adequate to the extent that it allows us to achieve the goals that are subjectively important for us. Viewing ethics in this way is quite consistent with a consequentialist approach. Gould specifically mentions that he cannot see the truth in Kant's categorical imperative (1993: 50). Kant's notion of *hypothetical* imperative, though, does seem to Gould to be a wise ethical principle; which is more evidence of Gould's consequentialism. If one has feelings for others, which human beings most certainly do, then it is possible to be pushed to accept the argument of consistency that supports a consequentialist utilitarian ethic. Gould does not overtly endorse utilitarian ethics but, given what he has said about human nature and ethics, it is quite possible for him to do so. An ethical principle that both he and Mill overtly endorse is the golden rule: Gould sees it as a wise ethical rule that is ultimately premised on enlightened self-interest (1993: 50).

As for the principle of equality, if human beings choose to construct egalitarian societies in which the interests of each being are given the same weight, then human reciprocal altruism, the golden rule, as well as flexible human potential are scaffolds to support the principle of equality. An egalitarian sentiment is implied in Gould's frequent claim that *Homo sapiens* is one species among millions (1993: 48). Further, Gould seems confident that the inegalitarianism that was once justified through social Darwinism rested on a spurious model of evolution (2002: 913).

Gould's multilevel theory of selection seems to provide enough conceptual space for accommodating the basic building blocks required to frame a utilitarian ethics. In the unfolding of evolution, evolutionary individuals often develop emergent properties (2002: 620). The elements of utilitarian ethics easily fit among the emergent properties of evolutionary individuals. Under the efficacy leg of Gould's revised Darwinian tripod, ethical aspects of the human species such as the building blocks of utilitarian ethics are possibly adaptations: "Human social practices," Gould says, "are clearly adaptive" (Gould 1977: 256). If there are other internal formal constraints at work that account for why human beings have sentient abilities, are social creatures, and have the cognitive ability to assess how their actions impact the welfare of others, that would only support viewing utilitarian ethics in a framework of human nature that has been crafted through evolution.

Because evolutionary theory is discussed on a different timescale than everyday ethics, we cannot solely use evolutionary theory in order to make ordinary ethical decisions. Gould says himself that geological time is not the proper scale of his personal life (1993: 48). Similarly, Gould also cautions that evolutionary explanations of human behavior are usually insufficient and that the forces of culture are very strong in shaping human social practices. His model of evolution does acknowledge, however, that it makes sense to view human nature as somewhat stable, which would provide building blocks for ethics. Specifically, Gould's model of evolution is consistent with the utilitarian view that one of the most essential characteristics of human nature is that human beings are creatures driven by feelings, emotions, and sentiments; and these emotional capacities allow them to have genuine concern for others.

10.3d Haught and Utilitarian Ethics

In terms of the elements of utilitarian ethics, what we can notice about Haught's aesthetic cosmological principle is that it is future oriented: human beings have the power to participate in bringing more complex beauty into the cosmos. Human actions are best when they contribute more beauty and allow creation to continue to unfold, and are worse when they either do not contribute to the beauty when it is in their power to do so, or they actively engage in diminishing the amount of beauty that already exists in the cosmos. In Haught's account we see the broad outlines of a consequentialist ethic: if we apply the aesthetic cosmological principle to human actions, then actions are better or worse depending upon the consequences they bring about. Utility, too, understood as *usefulness* in bringing about the desired end, which in this case is the amount of beauty in the cosmos, seems to be an apt concept for Haught's ethics.

Let us consider the other two ethical principles we have seen linked to utilitarian ethics, the golden rule and the principle of equality. Haught believes that theological engagements with science can benefit by taking more explicit notice of events such as the life and death of Jesus (2000: 109). Recall that the main question of this chapter concerns what evolution implies, if anything, about utilitarian ethics. Haught's model of theistic evolution says that those who are theorizing about science and religion can benefit from reflecting on the life of Jesus. Jesus obviously recommends the golden rule as an ethical principle, and as Mill said, the golden rule sums up the spirit of utilitarian ethics. But even more than accepting the golden rule as a dictum, what Haught has in mind is to notice how Jesus interacted with and respected those on the fringes of society. Jesus thought that those who occupied a low place in society should receive equal consideration as those in high society. It is a short step from the golden rule to the principle of equality, which says the interests of each human being should be given the same weight.

As we have seen, in the biblical tradition Haught sees the working out of the aesthetic cosmological principle in the parables of Jesus that include marginalized members of society (Haught 2000: 134). And Haught perceives an aesthetic dimension in the promotion of diversity and inclusiveness (Haught 2000: 134–35).

Haught's model of evolution is thus consistent with the main principles of utilitarian ethics. So committed is he to the aesthetic cosmological principle that he even welcomes contact with extraterrestrials as an opportunity to "enlarge our appreciation of God's love of diversity" (Haught 2003: 179). Extending feelings of sympathy to beings beyond one's tribe (a traditionally utilitarian program) would be consistent with the aesthetic cosmological principle.

Part of what is involved in the ongoing creation of complexity, nuance, and beauty is the generation of novelties. Millions of years have seen the generation of countless new forms of life, new species, and new ways of living. Here is another aspect of an evolving cosmos that has implications about human behavior and human ethics: when novelties are such an integral part of the evolutionary story, there is reason to think that new creations—such as cooperation in nature—are possible. Therefore, Haught's model of creative evolution has no difficulty in recognizing the utilitarian ethics of sentiment, sentience, and ethical altruism. In a human nature crafted by an inventive evolutionary process, the presence of social instincts and sympathetic feelings for others is not incredible.

A sticking point in harmonizing Haught's model of evolution with utilitarian ethics concerns the nature of the ultimate good. Classical utilitarianism says that pleasure is that which is most valuable (the doctrine of hedonism). If we depart from that doctrine slightly and agree at least that human happiness is the ultimate good, even this would fall short in Haught's view. Haught's evolutionary model is theistic. Humans surely have social sentiments, he'll say; but he disagrees with utilitarian theorists such as Hume and Mill who view these characteristics as accounting for the ultimate foundations of ethics. In Haught's model, ethics, like evolution, is ultimately of theistic origin.

We thus encounter a stumbling block when trying to harmonize Haught's model of evolution with utilitarian ethics. A naturalistic utilitarian ethics that tries to conjure the whole of ethics out of purely naturalistic materials without providing an explanation concerning the source of the cognitional act of "decision," which is driven by the imperative to "be responsible," will inevitably come up short.

Under Haught's model of evolution, a utilitarian view of human nature and ethics, although it correctly recognizes some things such as sentiments, consequences, a longing for goodness, the golden rule, and the importance of equality, does not, however, capture the whole story about ethics and human nature.

10.4 CONCLUSION

In this chapter we have seen how notions of utilitarian ethics fit in and make sense when placed in an evolutionary context. Dawkins makes very clear that in a genic selection process "Genes may maximize their selfish welfare at their level, by programming unselfish cooperation, or even self-sacrifice, by the organism at its level" (Dawkins 1995: 121–22). By Dawkins' own lights, then, unselfish cooperation or even self-sacrifice at the organismic level is compatible with selfishness at the genic level. Beyond the notions of unselfish cooperation, self-sacrifice, and issues of ethical altruism, though, other key elements of utilitarian ethics such as social sentiments, sentience, happiness, utility, consequentialism, and seeking after pleasure or seeking after well-being, also fit with Dawkins' genic selectionist view of human nature. Many of the same points Darwin made about ethics are still relevant and consistent with Dawkins' model as well as with contemporary versions of utilitarian ethics. Utilitarians claim that the evolutionary development of human beings actually helps to explain a bit more why a utilitarian ethic, which is based on sentiments, is plausible.

If we look at the ethical concepts and principles articulated by the utilitarian ethical tradition, utilitarian ethics still makes sense and remains quite serviceable even when looked at from various models of evolution. Utilitarian concepts and principles also fit with Haught's model of evolution. But for him they are ultimately insufficient in accounting for the native imperative to "be responsible" that Haught takes to be foundational. In the next chapter on care ethics we will examine the importance of a female perspective—maternal instincts specifically—for understanding ethics from an evolutionary perspective.

NOTES

1. Hume's thesis about the nature of ethics went contrary to the centuries-long tradition stretching from Plato, Aquinas, through Kant, that ethics is based on human rationality. But after Hume, the utilitarian ethical tradition has flourished (while all the while being severely criticized, of course).

2. Once one has acknowledged a sentimentalist approach to ethics, one can then go into a metaethical discussion and consider various characterizations of ethics such as metaethical noncognitivism, metaethical emotivism, metaethical expressivism,

metaethical quasi-realism, etc. (see chapter 3). In this chapter, though, I will focus strictly on utilitarian ethics as a normative ethical theory.

3. Not only has this egalitarianism pushed utilitarians in the nineteenth century to take up abolitionist and women's movements, but it prompted some utilitarians in the twentieth century to take up a nonanthropocentric agenda and extend the principle of equality to take into account the welfare of nonhuman animals. For, the moral community consists of all beings that are capable of happiness, pain, and suffering. Thus, sentience again is a relevant utilitarian concept (Bradie 1999: 32). One of the unique aspects of utilitarian ethics as compared with other ethical theories and traditions, is that utilitarians have pointed out that if we are committed to an egalitarianism ultimately based on a happiness principle, and happiness is simply defined in terms of pleasure and pain, then some nonhumans (those who experience pleasure and pain) should be regarded as equal to humans. Although most traditional ethical theories and traditions are *anthropocentric*, utilitarianism has been used to support a *nonanthropocentric* ethic, one that takes the focus off the human species and simply puts the emphasis on pleasures and pains; hence, on all sentient creatures, whether human or nonhuman. It is hard to deny that there are at least some nonhuman animals that experience pleasure and pain. Utilitarian ethics requires that we extend the moral community to include all sentient beings; it thereby expands the scope of ethics.

4. In a later work, Singer says he now realizes that his argument for the principle of equality cannot be squared with his metaethical position that denies that "normative claims can be true or false" (Singer 2011: 199). He proceeds to explore the possibility of defending the existence of "objective moral truths," in a manner recently suggested by Derek Parfit. He ends the book hopeful about such a project, but still tentative.

5. It seems, though, that if we employ Mill's distinction between higher and lower pleasures (1861: 8–11), Darwin's reservations do not damage the heart of utilitarian ethics. For, I may be giving up a lower pleasure when I make a sacrifice for my community, but in the process of sacrificing, I may experience a higher pleasure, for example, psychological satisfaction that I have the ability to help the community.

6. Dawkins explains his misgivings about popular misunderstandings of his views: "On the face of it, the Darwinian idea that evolution is driven by natural selection seems ill-suited to explain such goodness as we possess, or our feelings of morality, decency, empathy and pity . . . [we experience] wrenching compassion . . . powerful urge[s]. . . . Where does the Good Samaritan in us come from? Isn't goodness incompatible with the theory of the 'selfish gene'? No. This is a common misunderstanding of the theory. . . . I was mortified to read . . . that *The Selfish Gene* is the favourite book of Jeff Skilling, CEO of the infamous Enron Corporation. . . . I have tried to forestall similar misunderstandings in my new preface to the thirtieth-anniversary edition of *The Selfish Gene*" (Dawkins, 2006a).

7. Social Darwinism helped foster this way of thinking by suggesting that in unrestrained competition only the fittest will survive, thus it claimed that unrestrained competition is for the greater good (a seemingly utilitarian application). Notice how this popular line of reasoning assumes that *individual human beings* are competing against each other. But Dawkins says that the competition that is relevant to natural selection is *genic* competition. Under genic selection we cannot run the same

argument that so easily equates Darwinism with selfishness. Selfishness at the individual level may *not* be the best strategy for selfish genes. A better strategy for selfish genes might be to exist inside vehicles that are ethically altruistic. Ironically, then, although the use of the phrase "selfish gene" is often misunderstood in popular culture to mean that Dawkins denies the existence of ethical altruism or advocates selfishness, by shifting the focus of natural selection away from the individual and onto the gene he has provided conceptual space for deflating the popular association of Darwinism and selfishness.

Chapter 11

Evolution and Care Ethics

The primatologist Frans de Waal observes that "millions of years of selection for mothers who pay close attention to the needs of their offspring have promoted strong nurturing tendencies in mammalian females" (de Waal 1996).[1] Everyone knows what a mother bear will do to you if she sees you trying to harm her cubs, or even go near them! de Waal describes how "maternal protection is widespread and highly predictable" (de Waal 1996: 89–90). This maternal instinct is widespread due to natural selection, claims de Waal.

Even more importantly, though, for de Waal these strong and widespread nurturing tendencies have a broad impact on mammalian societies. He explains:

> Mammalian societies often are intimate groups in which everyone knows everyone else. Birth and death mark entry to and exit from the system: newborns are the center of attention, and attachments linger long after an individual's demise. The relationship between mother and offspring provides a blueprint for all other affiliations. Handicapped individuals may be fully accepted in the group. (de Waal 1996)[2]

In this chapter we will look at the normative theory known as care ethics, or ethics of care. Care ethicists have taken seriously the idea that "the relationship between mother and offspring provides a blueprint for all other affiliations," just as de Waal claims.

Defenders of care ethics argue that care ethics is an ethical approach distinctive from virtue ethics, natural law ethics, social contract ethics, deontological ethics, and utilitarian ethics. None of those normative approaches applies to personal relationships as well as care ethics does, they claim. Care ethicists point out that those ethical approaches work best in the public sphere, not in the realm of personal relationships.

Care ethics as a developed normative ethical theory has not been around very long.[3] "The ethics of care is only a few decades old," writes Virginia Held, an advocate of care ethics (Held 2006: 9). Care ethics was inspired by the work of psychologist Carol Gilligan (1982: 18), who exposed the male bias in Lawrence Kohlberg's theory of moral development.[4] In this chapter I sketch out care ethics as it has been developed by the moral philosophers Nel Noddings (1984, 2010) and Virginia Held (2006).

For Held, care ethics is

> based on universal experience: the experience of being cared for. It makes clear that persons need care or societies will not survive. . . . Empirically, in the context of the family, care has priority in the sense that without care, there will be no human beings. (Held 2006: 132–33)

Held argues that care ethics, while it takes the mother/child ethical relationship as fundamental, yet is much more than simply a motherly ethic, and she undertakes the task of applying care ethics to larger social and political issues that go far beyond the level of personal relationships.

The basic elements of care ethics include care, the one-caring, the cared-for, relationships, love, needs, and joy. After sketching out care ethics, the chapter then reviews how some theorists (Noddings, Rachels, MacIntyre, and Ruse) have characterized the relationship between care ethics and evolution. Lastly, the chapter views care ethics from the perspective of our four models of evolution.

11.1 THE CARE ETHICS OF NODDINGS AND HELD

For care ethics, living an ethical life involves caring for those with whom we are in close relationship. Care ethicists have worked out care theory to illuminate this central insight. Care theory is explicitly characterized as a normative ethical theory. Noddings says that she seeks to

> construct and defend a workable normative ethic . . . a prescriptive ethic should take adequate account of the way things are, including our inherited tendencies to care most about those closely related to us. (Noddings 2010: 27)

While Kant sought to identify *the* supreme principle of morality, from which can be derived all ethical duties, Noddings has more modest aspirations about care ethics. She writes that she is "exploring one significant source of morality—maternal instinct and the natural caring that develops from it" and does "not claim that it is the only source of morality" (Noddings 2010: 32).

She believes she can accomplish this without "denying other sources of morality" (Noddings 2010: 45).

Another distinctive feature of care ethics, says Noddings, is that it seeks to characterize the "full scope of moral life" (Noddings 2010: 36). "Like all normative ethics," she says,

> it is concerned with the moral justification of acts and decisions, but its emphasis is not on justification, and it does not concentrate on moments of decision. It is concerned with moral life considered whole, with what precedes and follows particular moral decisions. (Noddings 2010: 27–28)

To begin to develop this theory, care ethics must first answer the question, what is care? Noddings states that

> *caring* . . . involves attention, empathetic response, and a commitment to respond to legitimate needs. It is sometimes referred to as an attitude, but it is more than that; it is a set of dispositions to respond positively in interpersonal relations. Ethical caring involves a commitment. (Noddings 2010: 28)

Noddings distinguishes "natural caring" from "ethical caring." Natural caring is a response out of "love or natural inclination" (Noddings 1984: 5; 2010: 29). Natural caring comes "from our natural relatedness," and this "natural relatedness gives birth to love" (Noddings 1984: 43). We observe natural caring in everyday life, she says; it is a way of responding to each other (Noddings 2010: 17).

Ethical caring arises out of natural caring (1984: 4–5). "Natural caring is not a stage," however (Noddings 2010: 38). While natural caring flows from natural inclination, ethical caring takes conscious effort. She says that "our ethical caring depends upon both our past experience in natural caring and our conscious choice" (Noddings 1984: 157).

To genuinely care for another, says Noddings, another person is obviously implied: caring must be recognized as a caring *relation*. In care ethics,

> relation will be taken as ontologically basic and the caring relation as ethically basic . . . *relation* as ontologically basic simply means that we recognize human encounter and affective response as a basic fact of human existence. (Noddings 1984: 3–4; 2010: 33)

An important aspect of care ethics is its observation that humans are "relational beings" (Noddings 2010: 45). Care ethicists say this important insight about human nature is often overlooked, but it is part of human nature to be related and connected to others. Care ethicists stress and emphasize that human beings are relational beings that exist in many and various concrete relationships. Humans are born into families. Our relationships to family members are our first relationships. And without caring relationships, infants and small children will not survive into adulthood.

To be clearer when discussing the caring relation, Noddings employs the terminology of the "one-caring" and "cared-for." In her later work, she slightly modifies this language and designates the parties in a caring relation as carer and cared-for (Noddings 2010: 46).

An essential aspect of caring is attending to others' needs. "Care ethics is a needs-based approach," explains Noddings (Noddings 2010: 26). A carer is attentive to the expressed needs of the cared-for, yet those needs are not always verbally expressed (Noddings 2010: 47). The emphasis on needs makes care ethics distinct from other ethical approaches: "*Needs* are basic to a care-driven approach," she writes (Noddings 2010: 91).

Another important point about the caring relation is that genuine caring has *completion* in the other. The cared-for must in some way acknowledge the carer's effort (Noddings 2010: 48).[5]

Thus, people may talk about caring for things or caring for ideas, but those would not be genuine cases of caring, since the caring relation is not completed. "We may encounter a thing or an idea as 'Thou,'" writes Noddings, "but we do not receive feelings from it in response to our Thou-saying" (Noddings 1984: 161).

Care ethics is thought to bear a close resemblance to virtue ethics. But for Noddings, caring relations that have the quality of completion illustrate an important difference between these approaches (Noddings 2010: 49). Also, for care ethics, caring is more about the quality of relation, than about the virtues of the one-caring (Noddings 2010: 49).

Another important distinction that Noddings often makes is separating *caring* from *caregiving*. The elements discussed above that are parts of genuine caring may be absent in caregiving. Caregiving is an important form of labor that may or may not be accompanied by caring (Noddings 2010: 25).

Another interesting observation that Noddings makes about caring concerns the feeling that carers get when they are in a genuine caring relation. She calls it joy; not only is it a feeling, but it is a reward for the carer and it serves as encouragement and motivation. She describes how

> joy often accompanies a realization of our relatedness. It is the special affect that arises out of the receptivity of caring, and it represents a major reward for the one-caring. Feeling joy in relatedness . . . encourages growth in the ethical ideal. (Noddings 1984: 132)

"This joy arises out of an awareness of the caring relation," says Noddings (Noddings 1984: 138). Care ethics, "rooted as it is in relation, identifies *joy* as a basic human affect" (Noddings 1984: 6). When there is completion of a caring relation, the "Occurrence of joy is a manifestation of receptive consciousness" (Noddings 1984: 147).

I mentioned above how Noddings claims that care ethics takes seriously the inherited tendencies we have to care most about those closely related to us (Noddings 2010: 27). If so, then how about larger social and political issues? Does care ethics only apply to personal relationships? Caring relations as concrete relationships between individuals seem to defy extension to larger societal issues. Noddings says that we do not encounter people all over the world, so we cannot literally care for people all over the world (Noddings 2010: 80). When we donate to organizations that serve the poor, sick, and hungry we do not receive a response from them, so we should not think of donating as a caring relation (Noddings 2010: 81). For Noddings, this kind of situation points to the difference between individual obligation and collective obligation (2010: 80). If I as an individual am not in a caring relation with someone, then I cannot be said to be individually responsible for the needs of that person. But we can say that there is a collective responsibility to attend to the needs of the disadvantaged. Additionally, larger social and political issues commonly are justice issues, so Noddings advises that nations can adopt a care-driven concept of justice that acknowledges their collective responsibility (Noddings 2010: 89–98). Attending to needs is an essential aspect of a care-driven concept of justice, which she says contrasts with Kantian theories of justice that focus on rights as centrally important (Noddings 2010: 91).

Virginia Held is among the group of care ethicists who also work with applying care ethics to larger social and political issues.[6] She says that

> whereas the earlier forms of the ethics of care look especially at relations between persons within the family and among friends, the later developments make very clear that care ethics is not to be limited to the domain of the "personal." (Held 2006: 130)

She acknowledges that care at the personal level is the most compelling, yet that in itself does not mean that care ethics cannot be extended more broadly.

> We can see in the family and among friends the deepest and most compelling forms of care. But we can also see the relevance of the values of care and caring activities for the most comprehensive and global of moral concerns. (Held 2006: 132)

Held argues that care ethics has the "resources" to engage with and help with these larger issues and care ethics shows "promise" in doing so (Held 2006: 4). She claims that

> such caring relations are not limited to the personal contexts of family and friends. They can extend to fellow members of groups of various kinds, to fellow citizens, and beyond. We can, for instance, develop caring relations for persons who are suffering deprivation in distant parts of the globe. (Held 2006: 157)

In outlining how care ethics would envision a caring society, she sees care ethics as recommending ways of dealing with larger issues such as child care, education, and health care—obvious areas where a caring approach seems fitting—but also, the realm of law, the economy, and international affairs (2006, chapters 7–10).

11.2 CARE ETHICS AND EVOLUTION

Let us now look at care ethics from an evolutionary perspective. Nodding herself has recently taken up this task. She says that "evolutionary studies are having effects across the disciplines" (Noddings 2010: 38). "Much work is being done today on the evolution of morality," she observes (Noddings 2010: 1). And she welcomes an evolutionary perspective even if "future advances in evolutionary studies and genetics might produce other results we would prefer not to consider" (Noddings 2010: 5). She counsels that "we should be prepared for a variety of possibilities" (Noddings 2010: 5). Until recently, feminists did not trust what scientists had to say about the sexes (Vandermassen 2005: 1–2).

While Noddings thinks that current work on the evolution of morality is important, she is surprised that scholars in this area rarely look at female experience (2010: 1). In fact, research into the evolution of morality often entirely ignores female experience (Noddings 2010: 10). Vandermassen has also noted a lack of "attention to females in the historical process of evolutionary theorizing" (Vandermassen 2005: 60).

For Noddings herself, though, evolution plays an important part in understanding care ethics and the evolution of morality. Some of the very differences between males and females are due to evolution, she thinks. She takes an evolutionary perspective seriously, so regarding "the idea that women are essentially different from men," she thinks we can posit "fairly stable differences induced by evolutionary experience" (Noddings 2010: 24). Regarding these differences, her view is that "at least some can be traced to biological evolution" (Noddings 2010: 24).

She thinks that a biological perspective corroborates the important insight from care ethics that human beings are essentially "relational beings" (Noddings 2010: 45).

As we have seen, Noddings' version of care ethics highlights how care ethics differs from virtue ethics. She understands ethics to be more importantly about one's relations, less importantly about one's virtues. Here again, she thinks an evolutionary view supports her view, since the caring relation is more primary in an evolutionary perspective (Noddings 2010: 44).

To specifically see how evolution plays an important part in care ethics, we need to go back to Noddings' claim that one source of morality is the maternal instinct that became coupled with cognitive development and particular virtues (Noddings 2010: 45).

If the maternal instinct is one of the sources of morality, then what is the source of the maternal instinct? Evolution, obviously.

She thinks she can tell a plausible story about this. We only need to think about the earliest human mothers. In order for their infants to survive, early females needed to have particular characteristics and abilities, most important would be the ability to detect, "to read," when the vulnerable and defenseless infants needed something. About "the earliest human females," she reasons that

> a successful female—one who survived and nurtured more than one living infant—was aided by capacities associated with the maternal instinct. Maternal instinct prodded her to care about her infant's survival. She wanted it to live. To care successfully for it, however, she had to learn how "to read" the infant, understand the needs expressed, and have the resources to meet these needs. Is the infant hungry? Cold? In pain? In need of grooming? Females incapable of this basic reading would be unlikely to have surviving infants. (Noddings 2010: 11)

The maternal instinct of "reading" the infant needs more clarification. It involves feeling, but then also assessment and judgment of what to do. The infant expresses its needs through an expression of feelings, and the mother, by being open, and empathetic to feeling the same thing, detects what the infant is expressing (Noddings 2010: 12–13). We can get some sense of this transmission process if we think of the phenomenon of one infant merely hearing another infant cry, and then beginning to cry as well.

The maternal instinct explains the origin of natural caring. A maternal instinct manifests in love for the child, which gives rise to an attitude of caring (Noddings 2010: 35). As we have seen above, natural caring is caring done out of love and natural inclination. And ethical caring grows out of natural caring; but natural caring itself still has moral worth, argues Noddings. This is an area of disagreement that Noddings has with Kantian deontological ethics, because Kant's ethic says that an act performed out of a loving inclination has no moral worth, since it is not done out of principle. Mothers *do* act of inclination, though. But it doesn't seem plausible to Noddings that their acts have no moral worth because of this (Noddings 2010: 36).

Noddings holds that acts done out of love and inclination *do* have moral worth. She thinks that in close relationships, the most important acts do not arise from duty but from inclination (Noddings 2010: 36).

There can be more to caring than natural caring, though. In the following passage, she envisions the development of maternal instinct into ethical caring.

> The natural caring that has evolved from instinct gives rise to the processes already described in the discussion of empathy: learning to read the other, sympathy (feeling with), motivation displacement, and the reevaluation of what is read. But natural caring in the context of mothering also calls forth virtues such as responsibility, kindness, gentleness, patience, unselfishness, and cleanliness. Over time, immersion in the tasks of child-rearing has led beyond instinct to maternal thinking. (Noddings 2010: 73)

Maternal *thinking* develops over time, and "might promote further empathic development" (Noddings 2010: 57). With this cognitive development, the care can be regarded as ethical caring—it involves conscious effort and conscious choice. So, actually, with regard to Kant again, she says with him that natural caring and ethical caring are different, but she thinks that both of them have moral worth (Noddings 2010: 36). To see what Noddings is getting at with this distinction, we should take note that she does not argue that "an ethic of care evolves in a blindly biological way" (Noddings 2010: 1). She references Virginia Held's advice that when engaging in care ethics, "we should not lose sight of ourselves as moral subjects—as persons who think and make choices" (Noddings 2010: 1). Our intellectual abilities have also evolved and we can use them to reflect on ourselves (Noddings 2010: 1).

Both Noddings and Held maintain that care ethics goes beyond personal mother-child relations. Those foundational relations have evolutionary roots, but with developed empathy and cognition, caring relations can extend much further, to broader interpersonal relations (Noddings 2010: 58).

We have seen how care ethics is similar to virtue ethics. The well-known virtue theorist Alasdair MacIntyre has begun to take seriously some insights from care ethics and evolution. In *Dependent Rational Animals* (1999: x), MacIntyre admits he was wrong to leave human biology out of his earlier work in virtue ethics. He concedes:

> I now judge that I was in error thinking an ethics independent of biology could be possible. (MacIntyre 1999: x)[7]

If he had paid more attention to our biological nature, he thinks he would have seen that his prior account did not reckon adequately with "the nature and extent of human vulnerability and disability," which he now sees as a "central feature of human life" (MacIntyre 1999: x). Another central feature he identifies is "dependence."

> It is most often to others that we owe our survival, let alone our flourishing, as we encounter bodily illness and injury, inadequate nutrition, mental defect and disturbance, and human aggression and neglect. This dependence on particular others for protection and sustenance is most obvious in early childhood and in old age. (MacIntyre 1999: 1)

MacIntyre explicitly credits feminists such as Virginia Held for their work in highlighting how crucial these features are to human life. In moral philosophy generally, MacIntyre says that "an acknowledgement of anything like the full extent of that dependence and of the ways in which it stems from our vulnerability and afflictions is generally absent"; yet, he points out that

> feminist philosophers have recently done something to remedy this, not only by their understanding of the connections between blindness to and denigration of women and male attempts to ignore the facts of dependence, but also—I think here particularly of the work of Virginia Held—by their emphasis on the importance of the mother-child relationship as a paradigm for moral relationships. (MacIntyre 1999: 3)

For an ethical theory to be adequate, according to MacIntyre, it must describe what a moral life is for "beings who are biologically constituted as we are" (MacIntyre 1999: x). MacIntyre says that we "conceive of ourselves and imagine ourselves as other than animal," and we "become in consequence forgetful of our bodies and of how our thinking is the thinking of one species of animal," even though we have "a theoretical acknowledgment of the past evolutionary history of human beings" (MacIntyre 1999: 4–5).

As a virtue theorist, MacIntyre now acknowledges that humans need two different sets of virtues: the "virtues of independent rational agency," and also "the virtues of acknowledged dependence" (MacIntyre 1999: 8, 120). As we saw in chapter 6 on virtue ethics (§6.1), Aristotle said that friends are important for living a good life. But bearing in mind the importance of the virtues of acknowledged dependence, MacIntyre says we must rethink Aristotle's position.

> We are able to draw upon Aristotelianism to characterize that kind of friendship that we need, but we need more than Aristotle himself provides, because of Aristotle's reluctance to admit the extent to which our need for friendship is bound up with the sharing of our vulnerability and our wounds. (MacIntyre 1999: 164)

To which MacIntyre adds: "Yet what we should by now have learned from the virtues of acknowledged dependence is that this is a respect in which men need to become more like women" (MacIntyre 1999: 164).

11.3 MODELS OF EVOLUTION AND CARE ETHICS

11.3a Darwin and Care Ethics

Let us look at care ethics from the perspective of Darwin's model of evolution. Darwin's account of ethics based in his model of evolution seems quite congenial to care ethics. Like care ethicists, when Darwin looks at ethics he views sympathy, feelings, and social instincts as important ethical ingredients (Darwin 1871: 72). In discussing ethics, Darwin talks of emotions, such as sympathy, but he even talks of love: "It is certain," he says, "that associated animals have a feeling of love for each other" (Darwin 1871: 76).

For Noddings, all human beings feel the pain and joy of others to some degree; and she says for one who feels nothing, directly or by remembrance, we must prescribe re-education (1984: 92). For Darwin, likewise, sympathy is an essential aspect of the social instinct and someone who had no instinct of this kind would be tantamount to "an unnatural monster" (Darwin 1871: 90).

Even more significant, though, is that both Darwin and care ethicists view parental and filial affections as primary. Darwin thinks that the parental and filial affections lie at the base of the social instincts (1871: 81). Darwin estimates that these affections probably give rise to the pleasure we get from being in society (Darwin 1871: 80).

It is natural that we have more concern for those who are close to us. Care ethics is unique among normative theories because it accentuates the "parental and filial affections" more than any other normative approach.

Darwin notes that there are clear cases in which an individual who never risked his life for someone would then disregard the instinct of self-preservation and jump into a river to save a stranger (Darwin 1871: 87).

These kinds of actions seem to be "the simple result of the greater strength of the social or maternal instincts," Darwin thinks (Darwin 1871: 87). Passages like this align with what de Waal and the care theorists have said about the social instincts having their roots in the maternal instinct. As a normative ethical theory, not only do care ethicists concur in this genealogy of social instincts, but they maintain that actions in accord with these instincts are morally good. We feel that it is right to meet the needs of others, and care ethicists recommend that we trust this intuition as an ethical guide.

In chapter 9 on deontological ethics we saw that Ruse asserts that a Kantian indifferent universalization is in tension with a Darwinian view, because evolutionary biology predicts that we should have stronger feelings and a stronger sense of moral obligation for kin than strangers (1991a: 69). Regarding your own relatives and your children, Ruse says that kin selection insures that you have the emotions to give to them without expecting anything in return (Ruse 1998a: 239).

Ruse does not discuss care ethics by name, but in the way he describes a Darwinian view about feelings and obligations toward one's children, he sounds like he is specifically referring to a care ethics approach. A Darwinian says it makes perfect sense that you would feel that your children should come first (Ruse 1998a: 239).

The other kinds of feelings that care ethicists speak of—joy and love—would also make sense in a Darwinian model. Ruse would say that these feelings are ways that natural selection gets us to do what is in our biological interests. Recall how Noddings defines care:

> *Caring* . . . involves attention, empathetic response, and a commitment to respond to legitimate needs. It is sometimes referred to as an attitude, but it is more than that; it is a set of dispositions to respond positively in interpersonal relations. Ethical caring involves a commitment. (Noddings 2010: 28)

Every bit of this fits with a Darwinian model of evolution. It is in our biological interests to protect our kin, and the good feelings we experience when we are in caring relations with our kin serve to motivate us to preserve those caring relations.

11.3b Dawkins and Care Ethics

Turning now to Dawkins' genic selection model, it appears that with his model of evolution, Dawkins could easily make sense of the view of human nature that underlies care ethics. About genes, Dawkins claims that

> genes survive down the ages only if they are good at building bodies that are good at living and reproducing in the particular way of life chosen by the species. (Dawkins 1995: 5)

Noddings and Held have argued that without care there would be no human beings. Genes prompt a caring instinct as a way to preserve human bodies that in turn will help to preserve and reproduce genes. The maternal instinct and maternal feelings make perfect sense in terms of genic selection. Humans are social animals and care is a genetically based behavioral strategy that works well for social animals. To this we can add that working well means that successful human females were the ones who were able to read the needs of their infants, as Noddings speculated. Thus, Noddings' claim that the maternal instinct is one of the original sources of morality would seem to be a reasonable hypothesis if we understand evolution in terms of genic selection.

When Dawkins talks about genes that build bodies, he also means that genes can build bodily behaviors. So in *The Extended Phenotype*, where he calls genes "active germ-line replicators," he writes:

> To the extent that active germ-line replicators benefit from the survival of the bodies in which they sit, we may expect to see adaptations that can be interpreted as for bodily survival. A large number of adaptations are of this type. To the extent that active germ-line replicators benefit from the survival of bodies *other* than those in which they sit, we may expect to see "altruism," parental care, etc. (Dawkins 1982: 84–85)

In another passage from that book, Dawkins reminds us to be careful in how we describe behaviors that are genetically based. About sterile worker insects, for example, it is not correct to say that "sterile worker insects care for *other workers* because they share genes with them;" but rather, "workers care for their *reproductive* siblings who carry germ-line copies of the caring genes" (Dawkins 1982: 85). In talking of humans, then, it would be genetically predictable that mothers would care for their reproductive offspring, which carry the germ-line copies. (And Dawkins is of course using the word "care" in a looser sense than care theorists do.)

Noddings has claimed that theorists working with the evolution of morality seldom incorporate the female experience. Dawkins has tried to look at things from a mother's point of view (genetically speaking) (1989: chapters 7 & 8). With regard to genic selection, Dawkins has written that "the species with which we are most familiar—mammals and birds—tend to be great carers" (Dawkins 1989: 109).[8] There is an important distinction between "child-bearing and child-caring," Dawkins argues (Dawkins 1989: 109). Caring and bearing compete with each other he says, and speaking genetically, to achieve an evolutionarily stable strategy, there would have to be a mixed strategy of bearing and caring, since all caring and no bearing would not bring forth any new individuals from that gene line. And at the other extreme, mothers "who have too many children are penalized . . . because fewer of their children survive" (Dawkins 1989: 117).

Additionally, Noddings has written how for the caring relation to be completed, a response from the cared-for is required. Dawkins frames this in genetic terms, saying that

> it is of advantage to a parent to know when a baby is happy, and it is a good thing for a baby to be able to tell its parents when it is happy. Signals like purring and smiling may have been selected because they enable parents to learn which of their actions are most beneficial to their children. The sight of her child

> smiling, or the sound of her kitten purring, is rewarding to a mother. (Dawkins 1989: 138–39)

Another interesting area of agreement with Dawkins' model and Noddings care ethics concerns naturalism and supernaturalism. Dawkins offers a purely naturalistic account of evolution, and Noddings too, also frames care ethics naturalistically; and they both reject supernaturalism (Noddings 2010: 1–2).

An interesting point about this is that not only is Dawkins a naturalist but he has written about how a supernaturalist approach is objectionable for many reasons (2006a). Noddings has stated that supernaturalism has had a pernicious effect on women particularly. Even though care ethics does not require a supernatural dimension, she perceives that women have been susceptible to it, and have even upheld religious institutions that have oppressed them (Noddings 2010: 6–7).

11.3c Gould and Care Ethics

Gould's model of evolution is also naturalistic, but compared to Dawkins he is more tolerant of those who take a supernaturalist approach, because Gould thinks that although science and religion conflict, they are not necessarily incompatible, because they are nonoverlapping magisteria (NOMA). Gould advocates "respectful noninterference" between science and religion (Gould 1999: 5).

With regard to care ethics specifically, there are several aspects of Gould's evolutionary model that fit well with what care theorists say about human nature. The first is Gould's point that "weapons of survival need not be claws and teeth," that "patterns of reproduction may serve as well" (Gould 1977: 102). Care ethicists such as Held emphasize that without care there would be no human beings. Care is an essential ingredient for human survival; evolutionarily, newborn infants without access to maternal care simply will not survive. Gould's model can thus support care theorists in their claims about the biological importance of caring behavior. Further, under Gould's model, since care is a mainstay of the human social animal's strategy for survival, care can be thought of as a human trait that is in stasis.

Also, maternal care is a basic altruistic tendency. Gould's theory of hierarchical selection can accommodate the insight that kin selection is responsible for altruistic tendencies such as the maternal instinct (1977: 266). Gould's hierarchical theory of selection—which pictures selection proceeding on various levels—has no difficulty incorporating kin selection. Just like Dawkins, with Gould's model we can view the origins of altruistic feelings and desires as stemming from natural selection.

A point that Gould specifically makes about ethics also fits with how care theorists approach ethics. As we have seen in previous chapters, Gould views the Kantian categorical imperative that frames ethics in terms of a moral law that is absolute and unconditional as too unrealistic for creatures like us. In this chapter we have seen several examples of where Noddings points out that a care ethics approach importantly differs from a Kantian deontological approach. And these contrasts do trace back to the kind of moral principle the categorical imperative is. Noddings would be sympathetic with Gould's observation.

On the other hand, Gould may be suspicious of Noddings' acquiescence with what scientists have had to say about evolution and human nature. Gould is usually suspicious of how evolutionary theorizing is applied to human beings, because human behavior can be so flexible and unpredictable. Gould shares this view with earlier feminists who were less trusting about scientific claims regarding the sexes than Noddings is. Gould says it is very difficult to identify "a direct genetic basis for our most fundamental traits," and he may apply his point to the traits that care theorists have been highlighting about human nature, namely natural caring and humans as relational (Gould 1977: 238).

Rather than genetically fixed particular traits, folks like Gould see human flexibility as most important for human nature. It would seem that part of the care theorist's case about human nature is not about flexibility, but about the traits of natural caring, relationships, and maternal feelings and instincts. It is unclear whether Gould could really see natural caring and maternal care as having no genetic basis and see human nature as being so flexible that care is not necessary. Noddings is open to the possibility that future advances in science might yield results women would not prefer (Noddings 2010: 5). Even if the evolutionary story about care may turn out to be different, Noddings could still defend care ethics as a normative theory.

11.3d Haught and Care Ethics

Lastly, this brings us to Haught, who, in addition to accepting the science of evolution—evolution by natural selection—also defends a theistic evolution. As we have seen in previous chapters, Haught's overarching theistic model of evolution involves principles such as the narrative cosmological principle and the aesthetic cosmological principle. So the question in this section is what care ethics looks like from the perspective of Haught's theory of evolution. To begin to answer this question, we should first get clear on how Haught understands three concepts that are fundamental to care ethics: care, love, and relationships.

For Haught, if science shows that the origin of care as we know it today derives from a maternal instinct, which itself derives from the process of natural selection, then that would be fine with Haught, since he accepts evolutionary scientific accounts of human origins. But for Haught there is always a deeper source than science can penetrate, since science itself cannot account for "ultimates" (Haught 2010: 35).

Regarding care, Haught argues that there is a deeper kind of care, a cosmic care, which he views as the "deepest kind of care conceivable," and this kind of care only makes sense in a theistic evolutionary view of reality. Haught writes that

> theology after Darwin may now suggest that the universe, understood as an adaptive process itself, evolves at all only because in the remote reaches of its endless depth, there beckons something like a *promise*. It is here that providence resides and beckons. Providence is not manipulative of nature, but instead a reservoir of possibilities gently proffered to the world throughout its creative advance. This nonintrusive realm of endless possibility is, I believe, identical with the deepest kind of care conceivable, that is, with what theology calls providence (from the word "provide"). Evolution does not destroy but confirms the religious intuition that there is everlasting care at the bottom of things. (Haught 2003: 143)

This cosmic care is actually divine care and eternal care, and it relates to Haught's aesthetic cosmological principle.

> Additionally, as it considers the large question of the possible "point" or purpose of the universe, theology may now reasonably conjecture that this same divine care takes into itself all of the suffering, discord, tragedy and enjoyment in biological evolution and cosmic unfolding, as well as in human history. If the permanence beneath the process is endowed with the character of care, as theology is entitled to assume, it is not beyond reason to trust that this eternal care could also transform local cosmic contradictions into a wider harmony of contrasts, that is, into an unfathomable depth of beauty, and that our own destiny beyond death admits of conscious enjoyment of this beauty as well. (Haught 2003: 158)

While many of the above points about the universe rely on a scientific account of an expanding universe (as well as theology, of course), Haught contends that the view that there is an "inexhaustible Mystery" that "eternally grounds, encompasses and cares for the cosmos" can be more easily seen by those who are familiar with "biblical wisdom" (Haught 2003: 65). Haught sees his model of theistic evolution as incorporating "the biblical idea of a God of promise" (Haught 2010: 51).

The Bible also understands God as a God of love, adds Haught. And God as love allows the world to self-create itself through the long process of evolution. This is just what we should expect from a God of love. For Haught, force and annihilation have no place in a loving relationship. Love is nurturing and compassionate, and love even allows independence of that which is loved (Haught 2000: 39–40).

For Haught, God's "love might take the form of a self withdrawal" (Haught 2000: 40). Thus, as we have seen with Haught's model of evolution, he argues that the wandering and self-creating evolutionary process fits with a conception of a God who lets the world be (Haught 2000: 40).

God, therefore, should be thought of as "persuasive love, rather than coercive love," since to "compel would be contrary to the very nature of love" (Haught 2000: 41). God is also seen as a caring and self-giving love, as in Christianity we see a "God who submits to crucifixion" (Haught 2000: 47).

This way of understanding divine love helps us to understand human love. According to Haught: "It is the nature of love, even at the human level, to refrain from coercive manipulation of others" (Haught 2000: 53).

In addition to viewing the concepts of care and love theistically, Haught views the concept of relationship in a theistic manner as well. The God he describes is a "humble, utterly relational God" (Haught 2003: 82). As the ground of all being, God is in relationship with all aspects of the universe; additionally, Christians see God as a relational being in itself—father, son, and Holy Spirit, three in one being. And just as divine care and divine love have human counterparts, the same is true with relationality. "To exist in a bodily way" itself, according to Haught, "is fundamentally relational," and he says that all of the sciences have shown this (Haught 2000: 161). Haught would thus agree with Noddings who said that relation is ontologically basic (2003: 129). Haught could also welcome Noddings' analysis of the caring relation as involving a carer and a cared-for, since this is not something he specifies, although it would fit with his model.

In sum, humans have care, love, and relationality, but for Haught those qualities are a reflection of realities much deeper than human experience: deeper than culture, than biology, than evolution by natural selection, to the inexhaustible depth, that is, God.

With regard to the specifically normative aspects of care ethics, Haught does not specify the personal dimension of ethics nor the maternal dimension of care ethics. Haught does specify that caring relations should be extended and widened, however. He writes about the whole earth community, and how those who are normally left out should be included. Doing so is consistent with his aesthetic cosmological principle (Haught 2000: 134).

Haught says that for Christians, Jesus is the model for widening relationships to include "those who seemed relationless," for example, sinners and outcasts (Haught 2000: 163).

As noted, Haught does not explicitly acknowledge the personal dimension of ethics, that "tendency to care most about those closely related," which care theorists emphasize. Haught may see this tendency as a genetically based care that does not perceive the larger meanings and dimensions of caring relations. Haught would point out that Noddings says her approach is concerned with the moral life considered whole, and he would encourage a broader vision of ethics. Noddings says that a normative theory must take adequate account of the way things are. Haught's understanding of "the way things are" is theistic evolution. Interestingly, Noddings states that her theory describes one source of morality and she leaves open the possibility of other sources. With that, she has left an opening for Haught to fill with his suggested theistic source for morality.

Care theorists would insist, though, that care theory is meant as an ethical approach that takes seriously concrete personal relationships, and Haught's model encourages us to view ethics too abstractly, which is a common drawback care theorists see with most ethical theories. de Waal, too, sees that ethical theories can be too abstract and therefore unworkable. He writes that the "idea of universal brotherhood is unrealistic," for example (de Waal 1996: 214). Additionally, care theorists may see Haught's cosmic approach as an overly speculative theory about all of history that causes us to lose sight of our distinctive individuality and our unique caring relations.[9] Specifically, care theorists may point out that the very notion of divine relationality is overly abstract and defies the concrete specifications they see as crucial to genuine caring relations.

Haught, though, from the perspective of theistic evolution contends that most "professional ethical reflection is still excessively anthropocentric, blind to the fact that the earth and our humanity are folded into an immense universe," yet "our lives can be ennobled, and our ethical action animated, by knowing that we and the earth have an important part to play in the much-larger cosmic and Christic drama of creation" (Haught 2010: 147–48).

11.4 CONCLUSION

Care ethics is a recent and distinctive normative ethic. It is distinctive not only in its emphasis on care and personal relationships but also in its being inspired by maternal instincts. We must be clear to point out that maternal instinct and natural caring differ from ethical caring, as Noddings has described these terms.

Noddings contends that making moral decisions using care ethics differs from making moral decisions with deontological ethics, utilitarian ethics, and virtue ethics (Noddings 2010: 78–79). She has examined care ethics and the account of human nature that care theorists integrate into their ethical theory, alongside evolutionary thinking. Noddings finds that an evolutionary perspective supports her care theory. When sketching a picture of how care ethics can be traced to evolutionary roots Noddings does not specify exactly how she understands evolution, only that evolution is a scientifically based framework. In this chapter we have viewed care ethics from four models of evolution, Darwin's, Dawkins', Gould's, and Haught's, and found that overall there is indeed evolutionary support available for care theory. The fundamental elements of care ethics sit well with various models of evolution, even Haught's theistic model, though with important modifications being necessary. But it is doubtful whether care ethicists such as Noddings and Held would find Haught's evolutionary model helpful, not only because of its supernatural aspect but because Haught's approach would seem to revert ethics back to an impersonal ethic, due to its cosmic orientation.

NOTES

1. Stated under a photograph of a bonobo with her daughter, in a Photo Essay on "Closeness," between pp. 24–25.

2. Stated above a photograph of a Japanese monkey born without hands and feet, in a Photo Essay on "Closeness," between pp. 24–25.

3. It is true that Aristotle observed maternal care and love and mentions it in several places when writing about ethics (Aristotle 337 BCE), yet I wouldn't consider Aristotle's ethical approach a full-blown theory of care ethics.

4. See §8.3d, where I briefly discuss Kohlberg's theory. Vandermassen points out that Gilligan's critique of Kohlberg is well-grounded. "It is less clear; however," she writes, "whether Gilligan's equally influential study was more scientifically rigid than Kohlberg's" (2005: 28). Manning (1992) is another advocate of care ethics who also has reservations about Gilligan's research (1992, chapter 3).

5. To some critics, though, having the moral value dependent on uncontrollable circumstances, that is, whether the cared-for *receives* the caring, seems problematic. Michael Slote recommends to "drop the assumption from any account of the ethic of caring" (2001: 31).

6. Another member of this group of care theorists is Michael Slote. He proposes "*balanced caring*," a suggestion for balancing intimate caring with humanitarian caring (2001: 66–76).

7. Arnhart is supportive of MacIntyre's change of mind and offers argument to support MacIntyre's new approach (Arnhart 2001: 9).

8. Care theorists probably do not especially like Dawkins' way of describing a mother: "I am treating a mother as a machine programmed to do everything in its power to propagate copies of the genes which ride inside it" (Dawkins 1989: 123). For Noddings, being a mother is about occupying a particular kind of caring relation: "Whatever I do in life," she writes, "whomever I meet, I am first and always one-caring or one cared-for. I do not 'assume roles' unless I become an actor. 'Mother' is not a role; 'teacher' is not a role. When I became a mother, I entered a very

special relation—possibly the prototypical caring relation. When I became a teacher, I also entered a very special—and more specialized—caring relation" (Noddings 1984: 175).

9. Kierkegaard (1813–1855) is regarded as the father of existentialism; he asserted that Hegel's (1770–1831) speculative theory about all of history was a monumental abstraction that left an individually existing human being feeling meaningless and lost in an empty mansion.

Part III

EVOLUTION AND ETHICS

Chapter 12

Conclusion

In chapter 6 on virtue ethics we looked at Haidt and Joseph's contributions to evolution and virtue ethics. We saw how they view moral intuitions rooted in five separate domains of adaptive challenges (harm/care, fairness/reciprocity, in-group/loyalty, authority/respect, and purity/sanctity) (2007: 385). In their work they suggest an interesting analogy. They say that we can think of the moral intuitions of these areas as "taste buds" of the moral sense (2007: 385). Just as there are taste buds for sweet, sour, bitter, saltiness, etc., our moral intuitions, too, have a range of areas to which they apply. Further, just as our taste buds today can tell us about our ancestors—that they ate fruit and meat—our moral intuitions that correspond to the five domains of adaptive challenges can also tell us about our ancestors—that they were finely tuned for rearing children, helping kin, selectively cooperating, forming in-groups, etc.

Both our moral sense and our sense of taste contain diverse elements. And for both, we can perceive a meaningful unity in a diverse plurality. Each taste bud individually corresponds with a particular kind of taste. Yet we can also think of our sense of taste as a whole, perhaps a whole that is greater than the sum of its parts.

We have seen in the previous chapters how different theories emphasize and develop different aspects of ethics.[1] Various normative ethical theories highlight some aspects of ethics, and various metaethical theories highlight yet different aspects of ethics. These two levels of theories are distinct—metaethical theories will not give us practical guidance, and we cannot grasp the ultimate nature of ethics from one practical ethical concept or ethical principle. The ethical approaches we have reviewed and examined comprise a sum of the diverse parts of ethics. And, most relevant to the current research project, many of the following ethical theorists couple their ethical theory with an evolutionary perspective, for example, Mackie—error theory; Blackburn and Gibbard—expressivism; Wong—relativism; Railton—realism; Casebeer—virtue ethics; Arnhart—natural law ethics; Kahane—social contract ethics; Singer—utilitarian ethics; Korsgaard—deontological ethics; and Noddings—care ethics. Bearing in mind the four large animating questions initially stated in §1.1, we may perceive a meaningful unity in this diverse plurality of ethical approaches, and a whole that is greater than the sum of these diverse parts.[2] In the next section I will summarize my overall findings about evolution and ethics by quickly reviewing how I have answered those main questions.

Those main questions were:

Q1. If human biological evolution is part of our worldview, how do commonplace notions of ethics fit in? More specifically, if we understand human nature as the product of evolution, then how do standard ethical theories fit in?

Q2. Do traditional ethical concepts, principles, theories, and traditions make sense when placed in an evolutionary framework?

Q3. Do we need to abandon traditional ethics?

Q4. Do we need to create ethics anew; do we need a new evolutionary ethics?

12.1 QUESTIONS ANSWERED—SOLUTION OFFERED

In this work I have sought to outline the extremely rich array of work being done with evolution and both metaethics and normative ethics. This survey has yielded a solution to the philosophical problem of the nature and status of ethics in the light of evolution (§1.1), since I have answered the four main foundational questions that are central to this philosophical problem.[3]

To answer these questions, I have looked at scientific, philosophical, ethical, and theological works. I have considered the work of biologists, zoologists, paleontologists, philosophers, theologians, psychologists, and political scientists. Insights drawn from many disciplines can contribute to our understanding of human nature and ethics. Interdisciplinary dialogue is important because the details uncovered by one discipline may only be obscurely perceived by a different discipline that operates with an alternative methodology.[4]

As I said in §1.1, in asking the question, "what does evolution imply about ethics?" we must be clear about what we mean by ethics. Do we mean ethics in a more general sense, regarding the ultimate nature and status of ethics? Do we mean ethics in a practical sense, regarding conduct?

12.1a Evolution and Metaethics

Q1. For error theorists, a naturalistic evolutionary perspective would help to make error theory more persuasive, because a naturalistic evolutionary perspective combined with error theory helps to explain why commonsense morality is in error. Expressivists also argue that a naturalistic evolutionary perspective puts their metaethical view in a good light and is supportive. A theistic evolutionary perspective, by contrast, does not help error theorists or expressivists make their case. Haught would use his model of theistic evolution to show the implausibility of error theory and expressivism. Theists, after all, view ethics as having an ultimately supernatural origin, so they do not see ethical objectivity as an error, and they view ethics as much deeper than human sentiments.

Moral relativism and moral realism can also be conjoined with an evolutionary perspective, and proponents of those views seek to accommodate the insights of evolutionary theories. Interestingly, though, most naturalistic and theistic evolutionists view moral relativism as implausible on evolutionary grounds: most naturalistic and theistic evolutionary models view ethics as having a deeper structure than relativists seem to admit. Moral realists, on the other hand, can accommodate the insights of both naturalistic and theistic evolutionary theories.

Q2. With error theory, traditional ethics "make sense" in the light of evolution, in that traditional ethics are even more understandable as full of error. Expressivism as a metaethical theory simply redescribes commonsense ethical notions; so expressivists see commonplace notions of ethics fitting into an evolutionary framework just fine, as long as we understand ethics as ultimately derived from attitudes/feelings/sentiments. Both moral relativists and moral realists also view commonplace notions of ethics as fitting into an evolutionary perspective.

Q3. None of the four metaethical theories we have examined recommend that we need to abandon traditional ethics in the light of evolutionary considerations. Even error theorists who claim that commonsense morality is an illusion, still maintain that we *need* this illusion. (Mackie says we need everyday ethics; Ruse says we must all play along in order for the illusion of ethics to work and for us to benefit from ethics; Joyce says useful fictions are needed).

Q4. None of the four metaethical theories we have examined recommend that we need to create a new ethic or an "evolutionary ethic." Error theorists and expressivists contend that their views set against an evolutionary background are all the more persuasive. Moral relativists and moral realists seek to accommodate an evolutionary perspective into their theories and do not observe the need to create ethics anew.

12.1b Evolution and Normative Ethics

Q1. All six normative ethical theories we have examined can integrate and accommodate the insights of evolutionary theories, whether naturalistic or theistic models. Versions of Aristotle and Nussbaum's virtue ethics; social contract ethics of Hobbes and Kahane; deontological ethics of Kant and Korsgaard; utilitarian ethics of Mill and Singer; and care ethics of Noddings and Held can all be conjoined with an evolutionary perspective on human origins.

Q2. Commonplace notions of all six normative ethical theories fit into an evolutionary perspective.

Q3. We do not need to abandon traditional normative ethical approaches in the light of evolution, granted that with natural law ethics we only end up with a natural law type ethic. And with social contract ethics we can expand the power of social contract thinking, by using broader evolutionary models.

Q4. None of the six normative ethical theories we have examined recommend that we must create a new ethic or an "evolutionary ethic." Virtue ethics, natural law ethics, social contract ethics, deontological ethics, utilitarian ethics, and care ethics can all accommodate an evolutionary perspective.

12.2 POTENTIAL OBJECTIONS

My answering some potential objections may help to clarify my findings. The following potential objections are arranged in no particular order.

PO1. I have not explained "how" morality evolved. This is not something I have set out to do. I have not tried to articulate *the* story of where ethics comes from.[5] My project involves looking at a range of prevailing ethical theories, and also looking at evolutionary theories, and then determining the implications of evolution upon prevailing ethical theories. The question of how morality evolved is not necessarily relevant to this project.

PO2. I am trying to reduce ethics to biology. Reducing ethics to biology is not my motive. I do think that ethics has important roots in our evolved biology, however. And I think that "reducing ethics" can mean different things in this context. Railton, for instance, whose metaethical view we considered in §5.1b, distinguishes between a reduction that eliminates and a reduction that vindicates. Bearing in mind this distinction may help to alleviate the worry that motivates *PO2*.

Even error theorists do not propose an eliminative reduction for ethics, meaning that once they view ethics as reduced, they do view ethics as having become dispensable (§12.1a Q3). As we have seen, Railton himself defends a reductive naturalistic moral realism and he argues for a vindicating reduction of moral properties to natural properties. So even with a professed *reductive* naturalist, moral properties are not thereby dispensed with, but are seen as special kinds of complex natural properties; thus, though Railton is a reductivist, he still considers his position a version of moral realism.

PO3. I am attempting to prove that ethics comes from evolution.[6] As I said above, ethics has important roots in our evolved biology; but it would seem to be an overstatement to claim that all roots of ethics are biological roots. Even the scientific evolutionists we have heard from admit the importance of culture in the development of ethics.[7] Further, though, the very phrase "ethics comes from evolution" is ambiguous. Let us consider Korsgaard's deontological ethics approach for a moment. For her, morality emerges from the capacity for normative self-government (§§9.0, 9.1). She claims there is nothing nonnatural or magical about normative self-government, it can derive from self-consciousness. As her fellow deontologist, Nagel, puts it, a particular intellectual capacity is required for morality, and that capacity is presumably an effect or side-effect of biological evolution (Nagel 1978: 203–4). So according to the Korsgaard/Nagel timeline of human evolutionary history, once human beings had evolved brains that could yield the intellectual capacity of reflective self-consciousness, then humans would have become capable of normative self-governance, and ethics in the strong deontological sense could be said to have become possible. The evolutionary process would have then provided the necessary conditions for morality, but not the sufficient conditions.[8] Now here's the ambiguity. What if I said that both Nagel and Korsgaard believe that "ethics comes from evolution"? In one sense, according to their timeline of human evolutionary history, that statement would be true, since they admit that an evolutionary process gave rise to self-consciousness, which gave rise to normative self-government, which gave rise to ethics. But in another sense, the statement seems to be false, since under their view, ethics does not come *directly* from evolution, for Korsgaard and Nagel want to say that ethics arises from the capacity for normative self-governance.

Another example of how the phrase is misleading can be seen if we consider Haught's view. What if I said that Haught believes that "ethics comes from evolution"? Again, the statement is ambiguous. If we mean to say that

Haught believes that ethics comes solely from naturalistic evolution, the statement would be false. But if we are meaning to say that Haught believes that ethics comes from theistic evolution, the statement would be true. Overall, to say "ethics comes from evolution" is misleading; hence it is more appropriate to simply assert that ethics has important roots in our evolved biology.

PO4. I am attempting to derive an ethics from evolutionary considerations. This is not what I have attempted to do. The ethical theories I have examined have their place on the landscape of moral philosophy, independently of what their background assumptions about evolution are; so the ten ethical theories I have examined are not *derived from* evolutionary considerations. In the late nineteenth century, Social Darwinists tried to derive an ethics from evolutionary considerations, and Social Darwinism is a project that can now be seen as ill conceived.[9]

PO5. By premising my inquiry on the landscape of moral philosophy, I am making unwarranted assumptions. It might be objected that by assuming the distinction between is and ought, the distinction between metaethics and normative ethics, and separating biological inquiry from normative inquiry, I am assuming a Kantian ethic and a Kantian metaphysics. The objection implies that I have characterized all normative ethics as Kantian and thereby disallowed a Darwinian explanation of morality.

I do not think that observing a gap between statements of fact (is) and statements of ethics (ought), or a distinction between metaethics and normative ethics, or separating biological inquiry from normative inquiry commits one to a Kantian ethics, much less a Kantian metaphysics.

As I have previously mentioned (§1.5), I take the is/ought dichotomy to be simply a logical distinction: is-statements function differently than ought-statements. Likewise, the distinction between normative ethics and metaethics is a logical one. They are different types of inquiry that ask different questions. A normative ethical theory must answer the question: What should I do? A metaethical theory must answer the question; What is the status of ethics? Normative ethics offers practical guidance about what constitutes ethical conduct.

So further, since we are agents, we must decide how to act, and we do this from a first-person perspective. Since metaethics engages with more general questions about the status and origins of ethics, it is a level of inquiry operating at more of a third-person perspective. Biological science, also, provides a third-person perspective, thus it can be distinguished from a normative first-person perspective ethics.

These seem to me to be minimalist assumptions, not uniquely Kantian assumptions, so I do not think I am characterizing all normative ethics as

Kantian ethics. Kant may observe these distinctions, and attempt to shore them up with an extravagant metaphysics, but a Kantian metaphysics is not required to draw these logical distinctions. As I have mentioned, all ethical theories have some kind of background metaphysical assumptions (§1.4), but those assumptions can usually be separated from the specifically ethical components. I think contemporary Kantian deontologists have done this, for example, and sought only to develop and defend the ethical components of Kantian deontology.

PO6. I am appealing to scientific theories to justify ethical theories. As mentioned, the ethical theories examined in this book have their place on the landscape of moral philosophy. These ethical theories retain their places on the playing field of ethics on the merits of how well they deal with answering various questions about the nature of ethics and the practical demands of ethics.[10] Moral theories are often evaluated based on various criteria, such as their consistency, their ability to yield moral verdicts, their applicability, their intuitive appeal, their internal support, and their external support.[11] Theories of biological and theistic evolution are not ethical theories in themselves; thus, they are regarded as "external" to ethics. In this book, then, as we have examined ten ethical theories from an evolutionary perspective and considered how they fit with evolutionary theories, we have been exploring the external support for each ethical theory. Such an investigation is not objectionable, but is but one aspect of how ethical theories are evaluated.[12] In short, it is reasonable to inquire whether and how a particular ethical theory fits with our contemporary understanding of the world.

PO7. I am trying to get the theory of evolution to serve as the foundation for ethics. The relationship between evolution and ethics is more complex than this statement conveys. Just as stating that "ethics comes from evolution" is misleading (*PO3* above), similarly, stating that "evolution is the foundation for ethics" is also misleading. In both cases, issues of ambiguity reveal the shortcomings of these statements. Once more, it is more appropriate to describe my project as trying to show that the foundations of ethics have roots in our evolved biology, an unobjectionable claim I think, given the terrain of ethics we have traversed in this book. For, such a claim can accurately apply even to the foundations of deontological ethics, as described in the context of *PO3*, as well as accurately apply to a theistic ethic premised on Haught's model of theistic evolution. Haught asserts that "[o]ur capacities to perceive, understand and know have come to birth from within the bowels of earth and evolution;" thus, Haught could agree that the foundations of ethics have roots in our evolved biology (Haught 2006: 95). This does not commit him to the claim that all roots of ethics are biological roots.

12.3 EVOLUTION AND THE FOUNDATIONS OF ETHICS

After studying and appraising this large area of vibrant research, I do not think we should take seriously those commonly heard off-the-cuff remarks about ethics, evolution, and Darwinism.

Dawkins says we live in "a universe of blind physical forces and genetic replication" (Dawkins 1995: 133). If so, we would still need a normative ethic, since we still need to make decisions about what to do and what kind of life to live. The normative question, "what should I do?" is not sufficiently answered with insights from evolutionary theory.[13] Evolutionary biology gives us an account of where we came from, not where we *should* go, or what we *should* do.[14]

The reasonable question to ask, then, is not, "what does biological evolution say that I should do?" The reasonable question to ask is, if evolution is true, what does it imply about ethical approaches such as virtue ethics, social contract ethics, utilitarian ethics, etc., which have been developed independently of evolutionary theory? And we can ask the same reasonable question about metaethical accounts that were developed independently of evolutionary theory. What does evolution imply about *those* accounts?

All ethical theories, whether normative or metaethical, usually make assumptions about human nature (§1.4). Broadly speaking, then, we have been reviewing how an account of an evolved human nature might impact various ethical approaches. After investigating ethical theories through the lens of (several models of) evolution, we see that a wide spectrum of standard ethical theories today actually assimilate an evolutionary perspective. What this seems to show is that whether normative or metaethical, ethics has roots in our evolved biology. The core of ethics runs deep in our species.[15]

As I said in the introduction (§1.4), our characterizing the ethical theories discussed in this book as "theories of evolutionary ethics," does not make good sense. Today, because of our understanding of the pervasiveness of biological (and cosmic) evolution—in terms of its mechanisms and its history—all standard ethical theories can be regarded as "evolutionary ethics." And this indicates that the term "evolutionary ethics" does not meaningfully define an exclusive and distinct kind of ethical theory. As we have seen in the previous chapters, it is more worthwhile to discuss the ethical materials and components themselves—for example, the aspects of virtue ethics, expressivist ethics, care ethics, etc.—and *then* view those materials from an evolutionary perspective. By employing an evolutionary perspective, we will then understand those ethical elements more deeply.

NOTES

1. The various ethical theories canvassed in this book provide some indication of the diversity in ethics. About aggression, E.O. Wilson has written that "like so many other forms of behavior and 'instinct,' aggression in any given species is actually an ill-defined array of different responses with separate controls in the nervous system. No fewer than seven categories can be distinguished" (Wilson 1978: 101). Analogously, perhaps we could say that ethics too has at least as many categories.

2. Wilson contends that the innate rules of moral reasoning differ from those that societies consciously use. He says that modern societies use a hodgepodge of moral reasoning that is "simply, a mess" (Wilson 1998: 254). Wilson seems to want to hold ethics to a monism rather than accept a healthy pluralism. This is puzzling, given that he gladly accepts that there are at least seven kinds of aggression (see note 1 above). Greene has also written that "our moral judgments are driven by a hodgepodge of emotional dispositions" (2008: 72). And Greene judges that these emotional dispositions "were shaped by a hodgepodge of evolutionary forces, both biological and cultural" (2008: 72). "Because of this," Greene claims, "it is exceedingly unlikely that there is any rationally coherent normative moral theory that can accommodate our moral intuitions" (2008: 72).

3. This is how I have framed the problem, that is, these are the big questions I chose to focus on. Perhaps other theorists may frame the problem differently.

4. In the *Origin of Species* (1859) Darwin makes his case about evolution and natural selection by drawing on different fields of study, ranging from animal breeding, embryology, comparative anatomy, geology, etc. (Stamos 2008: 233).

5. de Waal (2006) *does* try to explain "how morality evolved." But according to Kitcher (2006: 138–39), it seems that de Waal's story is not detailed enough.

6. This objection is somewhat inspired by Hare's worries in the article "Is There an Evolutionary Foundation for Morality?" (Hare 2004).

7. As Dennett puts it, "Since in the case of humans (and only humans) there is always *another* possible source of the adaptation in question—namely culture—one cannot *so readily* infer that there has been genetic evolution for the trait in question" (Dennett 1995: 485). E.O. Wilson also acknowledges this distinction and views "cultural mimicry" as different from "more structured forms of biological adaptation" (Wilson 1978: 33). In sweeping fashion, Wilson observes that "it seems safe to assume that the greater part of the changes that transpired in the interval from the hunter-gatherer life of 40,000 years ago to the first glimmerings of civilization in the Sumerian city states, and virtually all of the changes from Sumer to Europe, were created by cultural rather than genetic evolution" (Wilson 1978: 88).

8. This is how the moral philosopher Alan Gewirth puts it (1993: 242, 255).

9. See notes 5 and 8 in chapter 4; and note 7 in chapter 10.

10. See note 1 in chapter 6 for MacIntyre's explanation of why Aristotle's ethical theory, though initially formulated in ancient Greece, still has a place on the landscape of moral philosophy today.

11. These criteria are spelled out in Timmons (2013: 12–16).

12. And because evaluating an ethical theory's external support is only one aspect of the evaluation of an ethical theory, it makes sense to say that evolutionary biology, *in itself*, cannot assure us whether the nature of ethics should be understood as realist or irrealist, as Sturgeon puts it, since reaching such a difficult verdict depends "on details that a general Darwinian outlook leaves quite undetermined" (1992: 102).

13. As Nagel puts it: "This is an instance of the more general truth that the normative cannot be transcended by the descriptive" (1997: 244). "The question 'What should I do?'" Nagel writes, "like the question, 'What should I believe?' is always in order" (1997: 244).

14. As the biologist Jerry Coyne puts it: "How can you derive meaning, purpose, or ethics from evolution? You can't. Evolution is simply a theory about the process and patterns of life's diversification, not a grand philosophical scheme about the meaning of life. It can't tell us what to do, or how we should behave" (Coyne 2009: 225). The philosopher Rolston makes a similar point: "Science has evident survival value, teaching us how to gain benefits that we desire. But what ought we to desire? . . . Science describes what is (or was, or will be), not what ought to be. . . . After science, we still need help deciding what to value; what is right and wrong, good and evil; how to behave as we cope" (Rolston 1999: 161–62).

15. This is a line from Singer: "The core of ethics runs deep in our species and is common to human beings everywhere. . . . Nevertheless, some people resist the idea that this core has a biological basis which we have inherited from our pre-human ancestors" (Singer 1981: 27). As Greene puts it: "Nature doesn't leave to us to figure out that saving a drowning child is a good thing to do. Instead, it endows us with a powerful 'moral sense' that compels us to engage in this sort of behavior (under the right circumstances). In short, when Nature needs to get a behavioral job done, it does it with intuition and emotion whenever it can. Thus, from an evolutionary point of view it is no surprise that moral dispositions evolved, and it is no surprise that these dispositions are implemented emotionally" (Greene 2008: 60).

References

Abbott, Philip. 2005. *Political Thought in America*. Long Grove, IL: Waveland Press.

Altham, J.E.J. and Ross Harrison. (eds.) 1995. *World, Mind, and Ethics: Essays on the Ethical Philosophy of Bernard Williams*. New York: Cambridge University Press.

Annas, Julia. 2003. "Virtue Ethics and Social Psychology," *A Priori* 2: 20–34.

Aquinas, Thomas. [1270] 1988. "The Summa of Theology." In P.E. Sigmund, ed. *St. Thomas Aquinas on Politics and Ethics*. New York: W.W. Norton & Company, 48–52.

Arendt, Hannah. 1963. *Eichmann in Jerusalem*. New York: Penguin Books.

Aristotle. [350 BCE] 1954. "*Rhetoric*," trans. W.R. Roberts. In *Rhetoric and Poetics of Aristotle*. Toronto, Canada: Random House.

Aristotle. [337 BCE] 1999. *Nicomachean Ethics*, trans. T. Irwin, 2nd ed. Indianapolis, IN: Hackett Publishing Company.

Arnhart, Larry. 1998. *Darwinian Natural Right: The Biological Ethics of Human Nature*. Albany, NY: State University of New York Press.

Arnhart, Larry. 2001. "Thomistic Natural Law as Darwinian Natural Right." In Paul, Miller, & Paul, 1–33.

Arnhart, Larry. 2004a. "The Darwinian Moral Sense and Biblical Religion." In Clayton & Schloss, 204–20.

Arnhart, Larry. 2004b. "The Incest Taboo as Darwinian Natural Right." In Wolf & Durham 2004, 190–218.

Arnhart, Larry. 2008. "The Darwinian Biology of Thomistic Natural Law." http://darwinianconservatism.blogspot.com/2008/03/darwinian-biology-of-thomistic-natural.html

Augustine. [388] 2005. *The Morals of the Catholic Church*. Whitefish, MT: Kessinger Publishing.

Augustine. [397] 1997. *The Confessions*, trans. by M. Boulding. New York: Vintage Books.

Axelrod, Robert. 1984. *The Evolution of Cooperation*. Basic Books.

Ayer, A.J. [1936] 1952. *Language, Truth, and Logic*. New York: Dover Publications.

Bekoff, Marc. 2004. "Wild Justice and Fair Play: Cooperation, Forgiveness, and Morality in Animals." *Biology and Philosophy* 19: 489–520.

Bendall, D.S. (ed.) 1983. *Evolution from Molecules to Men*. Cambridge: Cambridge University Press.

Bentham, Jeremy. [1789] 1988. *The Principles of Morals and Legislation*. Amherst, NY: Prometheus Books.

Blackburn, Simon. [1971] 1993. "Moral Realism." In Blackburn, 111–29.

Blackburn, Simon. 1981. "Reply to McDowell: Rule-Following and Moral Realism." In S.H. Holtzman and C.M. Leich, eds. *Wittgenstien: To Follow a Rule*. Boston: Routledge, 163–87.

Blackburn, Simon. [1985] 1993. "Errors and the Phenomenology of Value." In Blackburn, 149–65.

Blackburn, Simon. [1988a] 1993. "How to Be an Ethical Anti-Realist." In Blackburn, 166–81.

Blackburn, Simon. [1988b] 1993. "Attitudes and Contents." In Blackburn, 182–97.

Blackburn, Simon. [1990] 1993. "Just Causes." In Blackburn, 198–209.

Blackburn, Simon. 1993. *Essays in Quasi-Realism*. New York: Oxford University Press.

Blackburn, Simon. 1995. *Social Coordination, Egoism, and Nature*: *Radhakrishnan Memorial Lecture*. Rashtrapati Nivas, Shimla: Indian Institute of Advanced Study.

Blackburn, Simon. 1996a. "Securing the Nots: Moral Epistemology for the Quasi-Realist." In Sinnott-Armstrong & Timmons, 82–100.

Blackburn, Simon. 1996b. "I Rather Think I Am a Darwinian." *Philosophy* 71: 605–16.

Blackburn, Simon. 1998. *Ruling Passions: A Theory of Practical Reasoning*. New York: Oxford University Press.

Blackburn, Simon. 2001. *Being Good*. New York: Oxford University Press.

Boehm, Christopher. 2000a. "Conflict and the Evolution of Social Control." In Katz 2000, 79–101.

Boehm, Christopher. 2000b. "The Origin of Morality as Social Control." In Katz 2000, 149–83.

Boss, Judith A. 1998. *Perspectives on Ethics*. Mountain View, CA: Mayfield Publishing.

Boyd, Craig A. 2004. "Thomistic Natural Law and the Limits of Evolutionary Psychology." In Clayton & Schloss, 221–37.

Boyd, Craig A. 2007. *Shared Morality: A Narrative Defense of Natural Law Ethics*. Ada, MI: Brazos Press.

Bradie, M. 1999. "The Moral Status of Animals in Eighteenth-Century British Philosophy." In Maienschein & Ruse, 32–51.

Brink, David. O. 1984. "Moral Realism and the Sceptical Arguments from Disagreement and Queerness." *Australasian Journal of Philosophy* 62: 111–25.

Brink, David. O. 1989. *Moral Realism and the Foundations of Ethics*. Cambridge: Cambridge University Press.

Brooks, Arthur C. 2014. "The Trick to Being More Virtuous." *The New York Times*, November 27.

Campbell, Richmond. 1984. "Sociobiology and the Possibility of Ethical Naturalism." In Copp & Zimmerman, 270–96.

Casebeer, William D. 2003. *Natural Ethical Facts: Evolution, Connectionism, and Moral Cognition.* Cambridge, MA: MIT Press.

Churchland, Paul M. 1996. "The Neural Representation of the Social World." In L. May, M. Friedman, & A. Clark, eds. *Mind and Morals: Essays on Ethics and Cognitive Science.* Cambridge, MA: MIT Press, 91–108.

Churchland, Paul M. [1998] 2001. "Toward a Cognitive Neurobiology of the Moral Virtues." In *The Foundations of Cognitive Science*, ed. Joao Branquinho. New York: Oxford University Press, 77–98.

Cicero. [52 BCE] 1991. *The Republic.* In W. Ebenstein & A.O. Ebenstein, eds. *Great Political Thinkers.* Philadelphia: Holt, Rinehart and Winston.

Clayton, Philip and Jeffrey Schloss. (eds.) 2004. *Evolution and Ethics: Human Morality in Biological & Religions Perspective.* Grand Rapids, MI: William B. Eerdmans Publishing.

Cobbe, Frances Power. 1871. "Darwinism in Morals." *Theological Review* 8: 167–92.

Collier, John and Michael Stingl. 2013. "Evolutionary Moral Realism." *Biological Theory* 7: 218–26.

Comte-Sponville, Andre. 2001. *A Small Treatise on the Great Virtues.* New York: Henry Holt & Company.

Copp, David. 2008. "Darwinian Skepticism about Moral Realism." *Philosophical Issues* 18: 186–206

Copp, D. and D. Zimmerman. (eds.) 1984. *Morality, Reason and Truth: New Essays on the Foundations of Ethics.* Totowa, NJ: Rowman & Allanheld.

Coyne, Jerry A. 2009. *Why Evolution is True.* New York, NY: Penguin Books.

Crisp, Roger. (ed.) 1996. *How Should One Live? Essays on the Virtues.* New York: Oxford University Press.

Crisp, Roger and Michael Slote. (eds.) 1997. *Virtue Ethics.* New York: Oxford University Press.

Darwall, Stephen. 1998. *Philosophical Ethics.* Boulder, CO: Westview Press.

Darwin, Charles. 1859. *The Origin of Species by Means of Natural Selection.* New York: The Modern Library.

Darwin, Charles. [1871] 1981. *The Descent of Man*, 1st ed. Princeton: Princeton University Press.

Darwin, Charles. [1874] 1995. "Moral Sense (Chapter 4 of *The Descent of Man, 2nd* ed.)." In Thompson, 41–76.

Dawkins, Richard. 1979. "Twelve Misunderstandings of Kin Selection." *Zeitschrift fur Tierpsychologie* 51: 184–200.

Dawkins, Richard. 1982. *The Extended Phenotype: The Gene as the Unit of Selection.* New York: Oxford University Press.

Dawkins, Richard. 1983. "Universal Darwinism." In Bendall, 403–25.

Dawkins, Richard. 1987. *The Blind Watchmaker.* New York: W.W. Norton.

Dawkins, Richard. 1989. *The Selfish Gene.* New edition. New York: Oxford University Press.

Dawkins, Richard. 1995. *River Out of Eden: A Darwinian View of Life.* New York: Basic Books.

Dawkins, Richard. 2001. "Sustainability Doesn't Come Naturally: A Darwinian Perspective on Values." www.environmentfoundation.net/richard-dawkins.htm.

Dawkins, Richard. 2005. "Atheists for Jesus." *Free Inquiry* 25, http://richarddawkins.net/article,20,Atheists-for-Jesus,Richard-Dawkins.

Dawkins, Richard. 2006a. *The God Delusion*. New York: Houghton Mifflin.

Dawkins, Richard. 2006b. *The Selfish Gene*. 30th anniversary edition. New York: Oxford University Press.

Deigh, John. 1995. "Empathy and Universalizability." *Ethics* 105: 743–63.

Dennett, Daniel C. 1988. *The Moral First Aid Manual: The Tanner Lectures on Human Values*. Delivered at the University of Michigan, November 1986, http://www.tannerlectures.utah.edu/lectures/documents/dennett88.pdf

Dennett, Daniel C. 1995. *Darwin's Dangerous Idea: Evolution and the Meanings of Life*. New York: Simon & Schuster.

Dennett, Daniel C. 2003. *Freedom Evolves*. New York: Penguin Books.

Dennett, Daniel C. 2006. *Breaking the Spell: Religion as a Natural Phenomenon*. New York: Penguin Books.

Dewey, John. [1898] 1993. "Evolution and Ethics." In Nitecki & Nitecki, 95–110.

Duska, R. and M. Whelan. 1975. *Moral Development: A Guide to Piaget and Kohlberg*. New York: Paulist Press.

Farber, Paul Lawrence. 1994. *The Temptations of Evolutionary Ethics*. Berkeley, CA: University of California Press.

Farber, Paul Lawrence. 1999. "French Evolutionary Ethics During the Third Republic: Jean de Lanessan." In Maienschein & Ruse, 84–97.

Flew, A.G.N. 1967. *Evolutionary Ethics*. New York: Macmillan.

Foot, Philippa. [1978] 1982. "Moral Relativism." In Krausz & Meiland, 152–66.

Frankena, William. [1957] 1976. "Ethical Naturalism Renovated." In K.E. Goodpaster, ed. *Perspectives on Morality: Essays by William K. Frankena*. Notre Dame: University of Notre Dame Press.

Gardiner, Stephen M. (ed.) 2005. *Virtue Ethics: Old and New*. Ithaca: Cornell University Press.

George, Robert P. (ed.) 1992. *Natural Law Theory: Contemporary Essays*. New York: Oxford University Press.

Gert, Bernard. 2004. *Common Morality: Deciding What to Do*. New York: Oxford University Press.

Gewirth, Alan. 1993. "How Ethical Is Evolutionary Ethics?" In Nitecki & Nitecki 1993: 241–56.

Gibbard, Allan. 1982. "Human Evolution and the Sense of Justice." In P.A. French, T.E. Uehling, Jr., & H.K. Wettstein, eds. *Midwest Studies in Philosophy VII*. Minneapolis, MN: University of Minnesota Press, 31–46.

Gibbard, Allan. 1990a. *Wise Choices, Apt Feelings: A Theory of Normative Judgment*. Cambridge, MA: Harvard University Press.

Gibbard, Allan. 1990b. "Norms, Discussion, and Ritual: Evolutionary Puzzles." *Ethics* 100: 787–802.

Gibbard, Allan. 2002. "The Reasons of a Living Being." *Proceedings and Addresses of the American Philosophical Association* 76: 49–60.

Gilligan, Carol. 1982. *In a Different Voice: Psychological Theory and Women's Development*. Cambridge, MA: Harvard University Press.
Godfrey-Smith, Peter. 1994. "A Modern History Theory of Functions." *Noûs* 28: 344–62.
Gomez-Lobo, Alfonso. 2001. *Morality and the Human Goods: An Introduction to Natural Law Ethics*. Washington, DC: Georgetown University Press.
Gould, Stephen Jay. 1977. *Ever Since Darwin: Reflections in Natural History*. New York: W.W. Norton & Co.
Gould, Stephen Jay. [1980] 1998. "Sociobiology and the Theory of Natural Selection." In Ruse 1998b, 253–65.
Gould, Stephen Jay. 1981. *The Mismeasure of Man*. New York: W.W. Norton & Co.
Gould, Stephen Jay. 1993. *Eight Little Piggies: Reflections in Natural History*. New York: W.W. Norton & Co.
Gould, Stephen Jay. 1999. *Rocks of Ages: Science and Religion in the Fullness of Life*. New York: Ballantine.
Gould, Stephen Jay. 2002. *The Structure of Evolutionary Theory*. Cambridge, MA: Belknap Press.
Greene, Joshua D. 2008. "The Secret Joke of Kant's Soul." In W. Sinnott-Armstrong, 35–79.
Gupta, Bina. (ed.) 2002. *Ethical Questions: East and West*. Lanham, MD: Rowman Littlefield.
Haidt, Jonathan and Craig Joseph. 2004. "Intuitive Ethics: How Innately Prepared Intuitions Generate Culturally Variable Virtues." *Daedalus* 133: 55–66.
Haidt, Jonathan and Craig Joseph. 2007. "The Moral Mind: How Five Sets of Innate Intuitions Guide the Development of Many Culture-Specific Virtues, and Perhaps Even Modules." In *The Innate Mind* (2007), eds. Peter Carruthers, Stephen Laurence, and Stephen Stich. Oxford University Press, 367–391.
Haldane, J. and C. Wright. (eds.) 1993. *Reality, Representation & Projection*. Oxford: Oxford University Press.
Hare, R.M. 1952. *The Language of Morals*. New York: Oxford University Press.
Hare, John. 2004. "Is There an Evolutionary Foundation for Human Morality?" In Clayton & Schloss, 187–203.
Harman, Gilbert. [1975] 2007. "Moral Relativism Defended." In Shafer-Landau & Cuneo, 84–92.
Harman, Gilbert. 1977. *The Nature of Morality*. New York, Oxford University Press.
Harman, Gilbert. 1984. "Is There a Single True Morality?" In Copp & Zimmerman, 27–48.
Harman, Gilbert and Judith Jarvis Thomson. 1996. *Moral Relativism and Moral Objectivity*. Cambridge, MA: Blackwell.
Harms, W.F. 2000. "Adaptation and Moral Realism." *Biology and Philosophy* 15: 699–712.
Haught, John F. 2000. *God After Darwin: A Theology of Evolution*. Boulder, CO: Westview Press.
Haught, John F. 2003. *Deeper than Darwin: The Prospect for Religion in the Age of Evolution*. Boulder, CO: Westview Press.

Haught, John F. 2006. *Is Nature Enough? Meaning and Truth in the Age of Science*. New York: Cambridge University Press.
Haught, John F. 2008. *God After Darwin: A Theology of Evolution*, 2nd ed. Boulder, CO: Westview Press.
Haught, John F. 2010. *Making Sense of Evolution: Darwin, God and the Drama of Life*. Louisville: Westminster John Knox Press.
Hefner, Philip. [1997] 1998. "Cloning as Quintessential Human Act." In Ruse 1998b, 352–56.
Held, Virginia. 2006. *The Ethics of Care: Personal, Political, and Global*. New York: Oxford University Press.
Hobbes, Thomas. [1651] 1962. *Leviathan*. M. Oakeshott, ed. New York: Collier Books.
Hofstadter, Richard. 1944. *Social Darwinism in American Thought*. Boston: The Beacon Press.
Hume, David. [1739] 1989. *A Treatise of Human Nature*. Oxford: Oxford University Press.
Hume, David. [1751] 1983. *An Enquiry Concerning the Principles of Morals*. Indianapolis, IN: Hackett Publishing Co.
Hurka, T. 2001. "The Common Structure of Virtue and Desert." *Ethics* 112: 6–31.
Hursthouse, Rosalind. [1991] 1997. "Virtue Theory and Abortion." In Crisp & Slote 217–38.
Hursthouse, Rosalind. 1996. "Normative Virtue Ethics." In Crisp, 19–36.
Huxley, Thomas H. [1894] 1995. "Evolution and Ethics." In Thompson 1995: 111–50.
Jacobs, Jonathan. 2001. *Choosing Character: Responsibility for Virtue & Vice*. Ithaca, NY: Cornell University Press.
James, Scott M. 2009. "The Caveman's Conscience: Evolution and Moral Realism." *Australasian Journal of Philosophy* 87: 215–33.
James, Scott M. and Peter Carruthers. 2008. "Evolution and the Possibility of Moral Realism." *Philosophy and Phenomenological Research* 77: 237–44.
John Paul II. 1997. "The Pope's Message on Evolution and Four Commentaries." *The Quarterly Review of Biology* 72: 381–406.
Joyce, Richard. 2000. "Darwinian Ethics and Error." *Biology and Philosophy* 15: 713–32.
Joyce, Richard. 2001. *The Myth of Morality*. New York: Cambridge University Press.
Joyce, Richard. 2006. *The Evolution of Morality*. Cambridge, MA: MIT Press.
Kagan, Shelly. 1998. *Normative Ethics*. Boulder, CO: Westview Press.
Kahane, Howard. 1995. *Contract Ethics*. Lanham, MD: Rowman & Littlefield.
Kain, Patrick. 2005. "Interpreting Kant's Theory of Divine Commands." *Kantian Review* 9: 128–49.
Kant, Immanuel. [1785] 1981. *Grounding for the Metaphysics of Morals*, trans. J.W. Ellington. Indianapolis, IN: Hackett Publishing Company.
Kant, Immanuel. [1790] 1987. *Critique of Judgment*, trans. W.S. Pluhar. Ellington. Indianapolis, IN: Hackett Publishing Company.
Katz, L.D. (ed.) 2000. *Evolutionary Origins of Morality: Cross-Disciplinary Perspectives*. Bowling Green, OH: Imprint Academic.

Kitcher, Philip. [1993] 1994. "Four Ways of 'Biologicizing' Ethics." In E. Sober, ed. *Conceptual Issues in Evolutionary Biology*, 2nd ed. Cambridge, MA: MIT Press, 439–50.

Kitcher, Philip. 2004. "Evolutionary Theory and the Social uses of Biology." *Biology and Philosophy* 19: 1–15.

Kitcher, Philip. 2006. "Ethics and Evolution: How to Get Here From There." In Macedo & Ober, 120–139.

Kolata, Gina. 2015. "Chinese Scientists Edit Genes of Human Embryos, Raising Concerns." *The New York Times*, April 23, 2015.

Korsgaard, Christine M. 1996. *The Sources of Normativity*. In Shafer-Landau & Cuneo 2007, 93–106.

Korsgaard, Christine M. 2006. "Morality and the Distinctiveness of Human Action." In Macedo & Ober, 98–119.

Krausz, M. and J.W. Meiland. (eds.) 1982. *Relativism: Cognitive and Moral*. Notre Dame, IN: University of Notre Dame Press.

Lemos, John. 2008. *Commonsense Darwinism: Evolution, Morality, and the Human Condition*. Chicago, IL: Open Court.

Levy, Neil. 2002. *Moral Relativism: A Short Introduction*. Oxford: Oneworld.

Levy, Neil. (ed.) 2010. *Evolutionary Ethics*. Burlington, VT: Ashgate.

Lewis, C.S. 1947. *Miracles: A Preliminary Study*. New York: Macmillan.

Locke, John. [1689] 1980. *Second Treatise of Government*. Indianapolis, IN: Hackett Publishing Co.

Lyons, David. [1976] 1982. "Ethical Relativism and the Problem of Incoherence." In Krausz & Meiland, 209–25.

Macedo, Stephen and Josiah Ober. (eds.) 2006. *Primates and Philosophers: How Morality Evolved*. Princeton, NJ: Princeton University Press.

MacIntyre, Alasdair. 1984. *After Virtue: A Study in Moral Theory*, 2nd ed. Notre Dame, IN: University of Notre Dame Press.

MacIntyre, Alasdair. 1999. *Dependent Rational Animals*. Chicago, IL: Open Court.

Mackie, J.L. 1977. *Ethics: Inventing Right and Wrong*. New York: Penguin Books.

Mackie, J.L. [1978] 1985. "The Law of the Jungle: Moral Alternatives and Principles of Evolution." In J. Mackie and P. Mackie, eds. *Persons and Values: Selected Papers Volume II*. New York: Oxford University Press, 120–31.

Mackie, J.L. [1982a] 1985. "Co-operation, Competition, and Moral Philosophy." In J. Mackie and P. Mackie, eds. *Persons and Values: Selected Papers Volume II*. New York: Oxford University Press, 152–69.

Mackie, J.L. [1982b] 1985. "Morality and the Retributive Emotions." In J. Mackie and P. Mackie, eds. *Persons and Values: Selected Papers Volume II*. New York: Oxford University Press, 206–19.

Mackie, J.L. 1982c. *The Miracle of Theism*. New York: Oxford University Press.

Maienschein, Jane and Michael Ruse. (eds.) 1999. *Biology and the Foundation of Ethics*. New York: Cambridge University Press.

Manning, Rita C. 1992. *Speaking from the Heart: A Feminist Perspective on Ethics*. Lanham, MD: Rowman & Littlefield.

Mill, John S. [1861] 1979. *Utilitarianism*. Indianapolis, IN: Hackett Publishing Co.

Miller, Christian B. 2002. "Rorty and Moral Relativism." *European Journal of Philosophy* 10: 354–74.
Mizzoni, John. 1995. "Moral Realism, Objective Values, and J.L. Mackie." *Auslegung: A Journal of Philosophy* 20: 11–24.
Mizzoni, John. 2005. "Darwinian Ethics and Moral Realism." In F.A. Adams, ed. *Ethical Issues for the Twenty-First Century*. Charlottesville, VA: Philosophy Documentation Center, 199–212.
Mizzoni, John. 2006. "Does the Fact of Human Evolution Undermine the Facts of Morality?" *International Journal of the Humanities* 2: 2287–90.
Mizzoni, John. 2009. *Ethics: The Basics*. Malden, MA: Wiley-Blackwell.
Mizzoni, John. 2013. "Birds Trust Their Wings, Sharks Their Teeth, and Humans Their Minds: A Critique of Haught's Critical Intelligence Argument Against Naturalism." *Philo: A Journal of Philosophy* 16: 145–52.
Moore, G.E. 1903. *Principia Ethica*. Cambridge: Cambridge University Press.
Moser, Paul K. and Thomas L. Carson. (eds.) 2001. *Moral Relativism: A Reader*. New York: Oxford University Press.
Munro, Donald J. 2005. *A Chinese Ethics for the New Century*. Hong Kong: The Chinese University Press.
Nagel, Thomas. 1970. *The Possibility of Altruism*. Princeton, NJ: Princeton University Press.
Nagel, Thomas. [1976]. "Moral Luck." In Nagel 1979a, 24–38.
Nagel, Thomas. 1978. "Ethics as an Autonomous Theoretical Subject." In Stent 1978, 196–205.
Nagel, Thomas. 1979a. *Mortal Questions*. NY: Cambridge University Press.
Nagel, Thomas. 1979b. "Ethics Without Biology." In Nagel 1979a, 142–46.
Nagel, Thomas. 1986. *The View from Nowhere*. New York, NY: Oxford University Press.
Nagel, Thomas. [1997]. *The Last Word*. New York: Oxford University Press. In Moser & Carson 2001, 241–56.
Nitecki, M.H. 1993. "Introduction." In Nitecki & Nitecki 1993.
Nitecki, M.H. and D.V. Nitecki. 1993. (eds.) *Evolutionary Ethics*. Albany: SUNY Press.
Noddings, Nel. 1984. *Caring: A Feminine Approach to Ethics & Moral Education*. Berkeley, CA: University of California Press.
Noddings, Nel. 2010. *The Maternal Factor: Two Paths to Morality*. Berkeley, CA: University of California Press.
Nussbaum, Martha. 1988. "Non-Relative Virtues: An Aristotelian Approach." *Midwest Studies in Philosophy*, vol. XIII, 32–53.
Nussbaum, Martha. 1995. "Aristotle on Human Nature and the Foundations of Ethics." In Altham & Harrison, 86–131.
Olberding, Amy. 2012. *Moral Exemplars in the* Analects: *The Good Person is That*. New York: Routledge.
Paul, E.F., Miller, F.D., and J. Paul. (eds.) 2001. *Natural Law and Modern Moral Philosophy*. Cambridge: Cambridge University Press.
Pence, Gregory E. [1998] 2007. "Will Cloning Harm People? In Rachels & Rachels, 114–25.

Pinker, Steven. 2011. *The Better Angels of Our Nature: Why Violence Has Declined.* New York, NY: Viking.

Pius XII. [1950] 1956. *Humani Generis*. In A. Fremantle, ed. *The Papal Encyclicals in Their Historical Context*. New York: Mentor Books, 283–88.

Plantinga, Alvin. 1993. *Warrant and Proper Function*. New York: Oxford University Press.

Pollack, Andrew. 2015. "Genetically Engineered Salmon Approved for Consumption." *The New York Times*, November 19, 2015.

Pope, Stephen J. 1994. *The Evolution of Altruism & the Ordering of Love*. Washington, DC: Georgetown University Press.

President's Council on Bioethics. 2002. "Human Cloning and Human Dignity: An Ethical Inquiry." www.bioethics.gov.

Prinz, Jesse. 2012. *Beyond Human Nature: How Culture and Experience Shape the Human Mind.* New York: W.W. Norton & Company.

Rachels, James. 1990. *Created from Animals: The Moral Implications of Darwinism.* New York: Oxford University Press.

Rachels, James. 1999. *The Elements of Moral Philosophy*. New York: McGraw-Hill, 3rd edition.

Rachels, James and Stuart Rachels. (eds.) 2007. *The Right Thing to Do: Basic Readings in Moral Philosophy*, 4th ed. New York: McGraw-Hill.

Railton, Peter. 1986a. "Moral Realism." *The Philosophical Review* 95: 163–207.

Railton, Peter. 1986b. "Facts and Values." *Philosophical Topics* 14: 5–31.

Railton, Peter. 1989. "Naturalism and Prescriptivity." *Social Philosophy & Policy* 7: 151–74.

Railton, Peter. 1991. "Moral Theory as a Moral Practice." *Noûs* 25: 185–90.

Railton, Peter. 1992a. "Pluralism, Determinacy, and Dilemma." *Ethics* 102: 720–42.

Railton, Peter. 1992b. With Stephen Darwall and Allan Gibbard. "Toward *Fin de siècle* Ethics: Some Trends." *The Philosophical Review* 101: 115–89.

Railton, Peter. 1992c. "Some Questions About the Justification of Morality." *Philosophical Perspectives* 6: 27–53.

Railton, Peter. 1993a. "What the Non-Cognitivist Helps Us to See the Naturalist Must Help Us to Explain." In Haldane & Wright, 279–300.

Railton, Peter. 1993b. "Reply to David Wiggins." In Haldane & Wright, 315–28.

Railton, Peter. 1996. "Moral Realism: Prospects and Problems." In Sinnott-Armstrong & Timmons 1996, 49–81.

Railton, Peter. [1998a] 2003. "Aesthetic Value, Moral Value, and the Ambitions of Naturalism." In Railton 2003, 85–130.

Railton, Peter. [1998b] 2003. "Red, Bitter, Good." In Railton, 131–47.

Railton, Peter. 2000. "Darwinian Building Blocks." In Katz, 55–60.

Railton, Peter. 2003. *Facts, Values, and Norms: Essays Toward a Morality of Consequence*. New York: Cambridge University Press.

Railton, Peter. 2010. "Moral Camouflage or Moral Monkeys?" *New York Times*, July 18.

Rawls, John. 1971. *A Theory of Justice*. Cambridge, MA: Belknap Press.

Rawls, John. 1993. *Political Liberalism*. New York: Columbia University Press.

Richards, R.J. [1986] 1995. "A Defence of Evolutionary Ethics." In Thompson, 249–91.

Richards, R.J.1993. "Birth, Death, and Resurrection of Evolutionary Ethics." In Nitecki & Nitecki, 113–31.

Richerson, Peter J. and Robert Boyd. 2004. "Darwinian Evolutionary Ethics: Between Patriotism and Sympathy." In Clayton & Schloss, 50–77.

Ridley, Matt. 1996. *The Origins of Virtue: Human Instincts and the Evolution of Cooperation*. New York: Penguin Books.

Rolston, Holmes, III. (ed.) 1995. *Biology, Ethics, and the Origins of Life*. Boston: Jones and Bartlett.

Rolston, Homes, III. 1999. *Genes, Genesis, and God: Values and Their Origins in Natural and Human History*. New York, NY: Cambridge University Press.

Rolston, Holmes, III. 2004. "The Good Samaritan and His Genes." In Clayton & Schloss, 238–52.

Rousseau, Jean-Jacques. [1762] 1987. *On the Social Contract*. Indianapolis, IN: Hackett Publishing.

Ruse, Michael. 1989. "Is Rape Wrong on Andromeda? An Introduction to Extraterrestrial Evolution, Science, and Morality." In M. Ruse, ed. *The Darwinian Paradigm: Essays on its History, Philosophy, and Religious Implications*. London and New York: Routledge, 209–46.

Ruse, Michael. 1991a. "Evolutionary Ethics and the Search for Predecessors: Kant, Hume, and All the Way Back to Aristotle?" In E.F. Paul, F.D. Miller, and J. Paul, eds. *Ethics, Politics, and Human Nature*. Cambridge, MA: Basil Blackwell, 59–85.

Ruse, Michael. 1991b. "The Significance of Evolution." In Singer, 500–10.

Ruse, Michael. 1993. "The New Evolutionary Ethics." In Nitecki & Nitecki, 133–62.

Ruse, Michael. 1998a. *Taking Darwin Seriously: A Naturalistic Approach to Philosophy*. Amherst, New York: Prometheus Books.

Ruse, Michael. (ed.) 1998b. *Philosophy of Biology*. Amherst, NY: Prometheus Books.

Ruse, Michael. 1999. "Evolutionary Ethics in the Twentieth Century: Julian Sorell Huxley and George Gaylord Simpson." In Maienschein & Ruse, 198–224.

Ruse, Michael. 2001. *Can a Darwinian Be a Christian? The Relationship between Science and Religion*. New York: Cambridge University Press.

Ruse, Michael. (ed.) 2009. *Philosophy After Darwin*. Princeton, NJ: Princeton University Press.

Ruse, Michael and Edward O. Wilson. 1986. "Moral Philosophy as Applied Science." *Philosophy* 61: 173–92.

Scanlon, T.M. 1998. *What We Owe to Each Other*. Cambridge, MA: Harvard University Press.

Schonborn, Christoph. 2005. "Finding Design in Nature." *The New York Times*. July 7, 2005.

Shafer-Landau, Russ. 2003. *Moral Realism: A Defence*. New York: Oxford University Press.

Shafer-Landau, Russ and Terence Cuneo. (eds.) 2007. *Foundations of Ethics*. Malden, MA: Blackwell.

Shone, Steve J. 2004. "Cultural Relativism and the Savage: The Alleged Inconsistency of William Graham Sumner." *American Journal of Economics and Sociology* 63: 697–715.

Sidgwick, H. [1874] 1981. *The Methods of Ethics*, 7th ed. Indianapolis: Hackett Publishing Company.

Singer, Peter. 1975. *Animal Liberation*. New York: Avon Books.

Singer, Peter. 1981. *The Expanding Circle: Ethics & Sociobiology*. New York: Farrar, Straus & Giroux.

Singer, Peter. (ed.) 1991. *A Companion to Ethics*. Cambridge: Blackwell.

Singer, Peter. (ed.) 1994. *Ethics*. New York: Oxford University Press.

Singer, Peter. 1999. *A Darwinian Left: Politics, Evolution and Cooperation*. New Haven, CT: Yale University Press.

Singer, Peter. 2011. *The Expanding Circle: Ethics, Evolution, and Moral Progress*. Princeton, NJ: Princeton University Press.

Sinnott-Armstrong, Walter. [2000] 2007. "Expressivism and Embedding." In Shafer-Landau & Cuneo, 485–94.

Sinnott-Armstrong, Walter. (ed.) 2008. *Moral Psychology Volume 3: The Neuroscience of Morality, Emotion, Brain Disorders, and Development*. Cambridge, MA: MIT Press.

Sinnott-Armstrong, W. and M. Timmons (eds.) 1996. *Moral Knowledge?* New York: Oxford University Press.

Skyrms, Brian. 1996. *Evolution of the Social Contract*. Cambridge: Cambridge University Press.

Skyrms, Brian. 2004. *The Stag Hunt and the Evolution of Social Structure*. Cambridge: Cambridge University Press.

Sloan, Phillip R. 1999. "From Natural Law to Evolutionary Ethics in Enlightenment French Natural History." In Maienschein & Ruse, 52–83.

Slote, Michael. 2001. *Morals from Motives*. New York: Oxford University Press.

Sober, Elliott and David Sloan Wilson. 1998. *Unto Others: The Evolution and Psychology of Unselfish Behavior*. Cambridge, MA: Harvard University Press.

Sophocles. [441 BCE] 1977. *The Oedipus Cycle: Oedipus Rex, Oedipus at Colonus, Antigone*, trans. Dudley Fitts and Robert Fitzgerald. New York: Harcourt.

Stace, W.T. 1937. *The Concept of Morals*. New York: Macmillan.

Stamos, David N. 2008. *Evolution and the Big Questions*. Malden, MA: Blackwell.

Stent, Gunther S. (ed.) 1978. *Morality as a Biological Phenomenon*. Berkeley: University of California Press.

Sterba, James P. 2005. *The Triumph of Practice Over Theory in Ethics*. New York: Oxford University Press.

Sterelny, Kim. 2001. *Dawkins vs. Gould: Survival of the Fittest*. Duxford, Cambridge: Icon Books.

Stevenson, Charles. 1944. *Ethics and Language*. New Haven, CT: Yale University Press.

Stove, David. 1994. "So You Think You Are a Darwinian?" *Philosophy* 69: 267–76.

Street, Sharon. 2006. "A Darwinian Dilemma for Realist Theories of Value." *Philosophical Studies* 127: 109–66.

Sturgeon, Nicholas L. 1984. "Moral Explanations." In Copp & Zimmerman, 49–78.

Sturgeon, Nicholas L. 1992. "Nonmoral Explanations." *Philosophical Perspectives* 6: 97–117.

Sturgeon, Nicholas L. 1995. "Critical Study of Wise Choices, Apt Feelings." *Noûs* 29: 402–24.

Sumner, William G. 1906. *Folkways*. Boston: Ginn & Company.

Teehan, John. 2003. "Kantian Ethics: After Darwin." *Zygon* 38: 49–60.

Thompson, Paul. (ed.) 1995. *Issues in Evolutionary Ethics*. Albany: SUNY Press.

Timmons, Mark. 2008. "Toward a Sentimentalist Deontology." In W. Sinnott-Armstrong 2008, 93–104.

Timmons, Mark. 2013. *Moral Theory*, 2nd ed. Lanham, MD: Rowman & Littlefield.

Trivers, Robert. [1971] 1994. "The Evolution of Reciprocal Altruism." In Singer, 78–88.

Vandermassen, Griet. 2005. *Who's Afraid of Charles Darwin?* Lanham, MD: Rowman & Littlefield.

van Hooft, Stan. 2006. *Understanding Virtue Ethics*. Chesham, Bucks: Acumen Publishing Limited.

de Waal, Frans. 1996. *Good Natured: The Origins of Right and Wrong in Humans and Other Animals*. Cambridge, MA: Harvard University Press.

de Waal, Frans. 2006. "Morality Evolved: Primate Social Instincts, Human Morality, and the Rise and Fall of 'Veneer Theory.'" In Macedo & Ober, 1–80.

Wenz, Peter S. 2009. *Beyond Red and Blue: How Twelve Political Philosophies Shape American Debates*. Cambridge, MA: MIT Press.

Williams, Bernard. [1976]. "Moral Luck." In Williams 1981, 20–39.

Williams, Bernard. [1980]. "Internal and External Reasons." In Williams 1981, 101–13.

Williams, Bernard. 1981. *Moral Luck: Philosophical Papers 1973–1980*. New York: Cambridge University Press.

Williams, Bernard. 1983. "Evolution, Ethics, and the Representation Problem." In Bendall, 555–66.

Williams, Bernard. 1995. "Replies." In Altham & Harrison, 185–224.

Williams, Bernard and J.J.C. Smart. 1973. *Utilitarianism: For & Against*. New York: Cambridge University Press.

Wilson, Edward O. 1978. *On Human Nature*. Cambridge, MA: Harvard University Press.

Wilson, Edward O. 1980. *Sociobiology: The Abridged Edition*. Cambridge, MA: Belknap Press.

Wilson, Edward O. 1998. *Consilience: The Unity of Knowledge*. New York: Alfred A. Knopf.

Wolf, Arthur P. and William H. Durham. (eds.) 2004. *Inbreeding, Incest, and the Incest Taboo: The State of Knowledge at the Turn of the Century*. Stanford, CA: Stanford University Press.

Wong, David B. 1991. "Relativism." In Singer, 442–50.

Wong, David B. 2006. *Natural Moralities: A Defense of Pluralistic Relativism*. New York: Oxford University Press.

Zagzebski, Linda. 1998. "The Virtues of God and the Foundations of Ethics." *Faith and Philosophy* 15: 538–53.

Zagzebski, Linda. 2010. "Exemplarist Virtue Theory." *Metaphilosophy* 41: 41–57.

Index

About the Author

John Mizzoni, Ph.D. is a professor of philosophy at Neumann University, which is located just outside Philadelphia, in Aston, PA. He writes in the areas of metaethics, evolution and ethics, and environmental ethics. His work has appeared in a wide variety of scholarly journals and he is the author of *Ethics: The Basics* (Wiley-Blackwell, 2009) and *Catholic and Franciscan Ethics: The Essentials* (Cognella, 2014), and editor of *G.E.M. Anscombe and the Catholic Intellectual Tradition* (Neumann University Press, 2016) and *G.E.M. Anscombe and Human Dignity* (Neumann University Press, 2016).

www.ingramcontent.com/pod-product-compliance
Ingram Content Group UK Ltd.
Pitfield, Milton Keynes, MK11 3LW, UK
UKHW041632031026
15587UKWH00001B/237

9 780739 199831